Machine Learning for Hackers

R and Jupyter Notebooks

Isabella Romeo

http://CalculusCastle.com

Machine Learning for Hackers: R and Jupyter Notebooks, Zeroth Edition, by Bella Romeo. Last Revised on: 8/11/18 (Version 0.1.2563)
Sherwood Forest Books, Los Angeles, CA, USA
ISBN-13: 978-0-9966860-4-4
ISBN-10: 0-9966860-4-5
© 2018 by B. E. Shapiro. All Rights Reserved. No part of this document may be reproduced, stored electronically, or transmitted by any means without prior written permission of the author.

Isabella Romeo is a pseudonym. Portions of this book may have in fact been written by Bella's human.

Portions of this text have been taken from the text *Scientific Computation: Python Hacking for Math Junkies*, by Bruce E. Shapiro. Reprinted with permission.

THIS DOCUMENT IS PROVIDED IN THE HOPE THAT IT WILL BE USEFUL BUT WITHOUT ANY WARRANTY, WITHOUT EVEN THE IMPLIED WARRANTY OF MERCHANTABILITY OR FITNESS FOR A PARTICULAR PURPOSE. THE DOCUMENT IS PROVIDED ON AN "AS IS" BASIS AND THE AUTHOR HAS NO OBLIGATIONS TO PROVIDE CORRECTIONS OR MODIFICATIONS. THE AUTHOR MAKES NO CLAIMS AS TO THE ACCURACY OF THIS DOCUMENT. IN NO EVENT SHALL THE AUTHOR BE LIABLE TO ANY PARTY FOR DIRECT, INDIRECT, SPECIAL, INCIDENTAL, OR CONSEQUENTIAL DAMAGES, INCLUDING LOST PROFITS, UNSATISFACTORY CLASS PERFORMANCE, POOR GRADES, CONFUSION, MISUNDERSTANDING, EMOTIONAL DISTURBANCE OR OTHER GENERAL MALAISE ARISING OUT OF THE USE OF THIS DOCUMENT OR ANY SOFTWARE DESCRIBED HEREIN, EVEN IF THE AUTHOR HAS BEEN ADVISED OF THE POSSIBILITY OF SUCH DAMAGE. IT MAY CONTAIN TYPOGRAPHICAL ERRORS. WHILE NO FACTUAL ERRORS ARE INTENDED THERE IS NO SURETY OF THEIR ABSENCE.

Early editions suck. Especially ones without editors. Don't expect this one to be any better. The POD model keeps prices low, but the reader becomes the editor. Please report any errors, omissions, or suggestions for improvements through the form at https://github.com/biomathman/machine-learning-in-R-book/issues, or by writing to the author directly at ibellaromeo@gmail.com. Bella will bark at all emails.

The only authorized distributor of the electronic version of this book is Gumroad. If you have obtained an electronic copy from any other source your copy is unauthorized and in violation of international copyright law.

Table of Contents

Contents iii
Preface v

I Getting Started   1
1 What is Machine Learning?   2
2 Jupyter Notebooks and R   4
3 Just Enough R   10

II Regression   32
4 Linear Regression   33
5 Polynomial Regression . .   46
6 Multilinear Regression . .   57
7 Nonlinear Regression . .   66
8 Backprop Networks . . .   70
9 Regression Trees   87

III Classification   99
10 Logistic Regression   100
11 Evaluating Binary Classification   107
12 Deep Learning with Keras   115
13 K-Nearest Neighbors . . .   127
14 Naive Bayes' Classifiers .   131
15 Discriminant Analyses . .   138
16 Principal Component Analysis   145
17 Support Vector Machines   158
18 Decision Trees   163
19 Bagging   169
20 Boosting   172
21 Random Forests   176
22 K-Means Clustering . . .   180
23 Hierarchical Clustering . .   193
24 DBScan   205

IV Other Random Topics   211
25 Self Organizing Maps . .   212
26 Hopfield Networks   220
27 Image Analysis   228
28 Afterward   247

V Back of the Book   249
A Review of Linear Algebra   250
General References   270
Index   272
About the Author   276

The Hacker's Codes

First Generation

- Computer access shall be unrestricted.
- Information shall be free.
- Judge others **only** by their acts.
- Computers shall produce beauty.
- Computers shall improve life.

Second Generation

- Do no evil.
- Protect data.
- Protect privacy.
- Conserve resources.
- Telecommunication shall be unrestricted.
- Share code.
- Strive to improve.
- Be prepared for cyber-attack.
- Always improve security.
- Always test and improve the system.

Preface

This is a book for hackers, programmers, engineers, scientists, and anyone else who wants to get down and dirty with machine learning (ML) but doesn't necessarily have the mathematical sophistication to learn a lot of advanced theory. Think of it as a lab manual in machine learning. The target audience is advanced community college students and lower division math, computer, and engineering students.

Figure 1.: Bella's late partner Romeo loved to collect bones. Bella and Romeo were inseparable for over seven years.

The chapters provide a survey of methods and techniques that I have found useful. None of the material is terribly deep, and a lot of subjects are skipped. The goal is to give you just enough information to get started at solving problems in ML using R. The chapters are relatively independent, so unlike most math or CS books, you don't have to start at page 1 to get to chapter 432. Just jump into whatever chapter you need. If something is needed from an earlier chapter, you will be sent there to find it (when necessary).

No previous experience with R is needed to read this book. You can learn R as you go along, though you might want to pick up a reference manual such as [ADLER]. This is not a book about R, so if you have never programmed before (or even if you have) you will very likely find the R language confusing. Keep in mind that the internet is your friend! Everything about R is thoroughly documented online and indexed by Google.

Jupyter notebooks for most chapters can be found at the companion website, https://github.com/biomathman/machine-learning-in-R-book. There is also a short form there where you can post feature requests or report typographical errors.

While there is a fair amount of mathematical content, its really just there to pique your interest. For the most part, if you just want to hack, you can skip all the math and go straight to the recipes. For those who want to delve deeper, I would suggest going to one of books in the short list in the back of the book. These books are chosen to span a variety of levels of mathematical sophistication ranging from (very little) through (undergraduate) to (advanced).

In the long run if you really want to learn about ML, you really should study the math. Since ML is a growing field, new methods are being published every day and you'll probably end up deep in the scientific literature if you want to be in the center of things. You'll need the math if you want to know how to implement these new ideas, figure out why they work, and why and why they fail so miserably at other times. For now, feel free just to hack and have fun.

Oh, and like that other guy said, thanks for all the fish!

Part I.
Getting Started

Just get a computer and start hacking.

1. What is Machine Learning?

Ask any two data scientists the question that this chapter asks and you will get two different answers. There is no satisfactory answer to the question "What is *machine learning?*"

The same can be said for the question "What is *Data Science?*"

The term *machine learning* is used to emcompass a wide variety of data analysis techniques. Many of the techniques were invented and in common practice long before the invention of computers. Some of these techniques are based on probability theory and resemble statistical analysis. Broadly categorized, *machine learning* encompasses techniques that will help user perform one or more of the following overlapping functions:

- Curve and model fitting, such as regression;
- Classification, such as grouping items into categories;
- Identification, such is figuring out which category an item fits into;
- Decision making, such as figuring how to schedule all the meetings in different conference rooms during a large convention;
- Recognizing patterns such as peoples faces, words, or handwriting;
- Prediction or extrapolation.

Oftentimes you will hear ML geeks break things down into *regression classification*. (In fact that's even how this book is organized.) But to some extent, many of these tasks involve aspects of both regression and classification. Methods that emphasize statistics are sometimes called *computational statistics*[1] or *statistical learning*.

The word learning is probably included in the expression *machine learning* because in order to do any of this, some sort of *model* is *fit* to a *data set*. The model must be learned. There are two general classes of learning:

> ▸ **Supervised Learning** - the model is fit based on examples of "correct" data. For example, given a list of student SAT scores and a flag that indicated whether or not they graduated or did not gradu-

---

[1] Although there are many areas of computational statistics that are unrelated to machine learning.

ate from a particular college, can the model predict the probability of a student (that was not listed in the original data set) eventually graduating as a function of their SAT score?

- **Unsupervised Learning** - the algorithm attempts to find structure in the data without being given a way to determine whether or not the solution it finds is correct. For example, given a Google map image, can it classify the pixels that belong to roads into a different category than pixels that do not belong to roads, without having the pixels labeled as "road" or "non-road" in advance?

Many times a single problem will involve aspects of both supervised and unsupervised learning and will require both regression and classification tasks.

The *data set must exist before the model is fit*, so machine learning is not a sub-discipline of mathematical modeling in its pure sense. Furthermore, it is not a sub-discipline of *data mining*, which is the process of exploring existing data sets. *Machine learning* is a collection of tools that is used in *data mining*.[2]

Finally neither *data mining* nor *machine learning* are sub-fields of statistics, because both are *data-driven*. That means that the data must exist before you start looking for anything in particular. *Statistics*, particularly *predictive statistics*, is hypothesis driven. In predictive statistcs, you must make a hypotheses *before any data is collected*. Many of the tools of *descriptive statistics*, however, are used to describe existing data sets.

---

[2]Other tools, for example, would be data base managers.

2. Jupyter Notebooks and R

The Jupyter notebook[1] is like a lab notebook for computer programmers. It lets you mix text (including fancy typeset text like latex, html, and markup), pictures and code in any order.(See figure 2.1.) There is even a collection of interactive widgets you can use.[2] For scientists, this means that you don't have to keep different parts of your program in different files, and you don't have to worry about what to type on the command line, since you can put the instructions right there in the notebook. You don't have to search through a folder of png or tiff files for your ouput, since they are also embedded in the file. It supports reproducibility since you can give the file to another user, and they can run the entire sequence of operations form top to bottom of your file (if you also give them the input data sets). Ideally a scientist would supply their Jupyter notebook as supplementary material along with a publication to a journal.

The name Jupyter is derived from the names of three computer languages it was originally designed to support: Julia, Python, and R. Originally developed as an outgrowth of the IPython project, there are kernels for nearly one hundred different computer languages[3] that can use the Jupyter notebook now, and the number seems to be growing almost daily. Installation still requires having Python on your computer, but you don't need to know or use Python to use Jupyter. Chances are that you already have some form of Python on your computer anyway.

Installation

To use R in the Jupyter notebook you must already have a working R installation. You must also have a working Python installation. Once Jupyter is installed, you also needed to install a program called the kernel, which connects Jupyter to R. An additional kernel must be installed for each computer language you will be using in Jupyter.

---

[1] See http://jupyter.org/documentation for full documentation.
[2] See http://jupyter.org/widgets.
[3] A list of kernels is available at https://github.com/jupyter/jupyter/wiki/Jupyter-kernels

Figure 2.1.: A sample Jupyter notebook. Notebooks can intermix text, code, and graphics in any order.

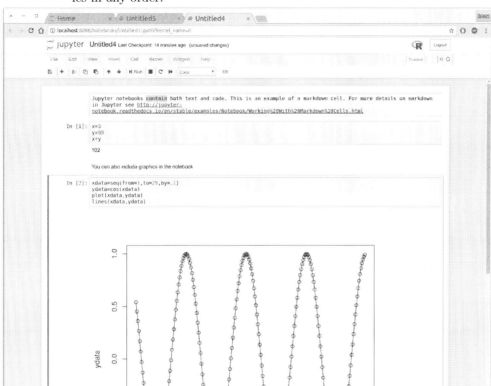

1. **Install R**

   R should be downloaded from CRAN[4]. CRAN is the Comprehensive R Archive Network. Follow the instructions to unpack the archive and install it on the CRAN web page.

   After installation, verify that it works by opening **terminal** in linux or MacOS, or **Command Prompt** in Windows, and typing **R** (a single upper case R) and then hit the [enter] key.

   You will know you are in the R console when it spits out a lot of copyright stuff and ends with a prompt "**>**". It will look something like this:

---

[4]At https://cran.r-project.org/

```
R version 3.4.4 (2018-03-15) -- "Someone to Lean On"
Copyright (C) 2018 The R Foundation for Statistical
Computing Platform: x86_64-pc-linux-gnu (64-bit)

R is free software and comes with ABSOLUTELY NO WARRANTY.
You are welcome to redistribute it under certain conditions.
Type 'license()' or 'licence()' for distribution details.

  Natural language support but running in an English locale

R is a collaborative project with many contributors.
Type 'contributors()' for more information and
'citation()' on how to cite R or R packages in publications.

Type 'demo()' for some demos, 'help()' for on-line help, or
'help.start()' for an HTML browser interface to help.
Type 'q()' to quit R.
>
```

Type **q()** to exit the program.

2. **Install Python 3**

   The latest version of Python can be installed from https://www.python.org/downloads/. Most Macs already have Python 3 installed. Linux users may prefer to use the latest version their repository. As of this writing, Python 3.6.5 was the lastest stable release, and all versions of Python 3.7 and above were experimental. Version 2.7.15 is also stable but there are differences and if you don't care you should use version 3.

   Some users prefer a commercial system like Anaconda Python because they are easy to install and have dashboards for opening the program. Anaconda has the advantage that it can be installed for a single user without administrator privileges on most Windows systems.

3. **Install Jupyter**

   You can install Jupyter using **pip** from the command line.[5] Open **terminal** in linux or MacOS, or **Command Prompt** in Windows, and

---
[5] See http://jupyter.org/install

type the following:

```
python3 -m pip install --upgrade pip
python3 -m pip install jupyter
```

If you are using Anaconda Python, you should skip this step.

4. **Install The R Kernel for Jupyter**

   To install the R Kernel[6] open up an R console (do not use a dashboard like **RStudio** for this). To open the R console, open a command line program and type **R**. You must do this from an R console that was started from the command line, not from an R conosole that was started from an R Application (such as the one that is installed on MacOS), becuase your installations will be lost otherwise.

   At the prompt type the following to install the necessary packages (Don't copy in the prompts!):

   ```
   > install.packages(c('repr', 'IRdisplay', 'evaluate',
   'crayon', 'pbdZMQ', 'devtools', 'uuid', 'digest'))
   > devtools::install_github('IRkernel/IRkernel')
   ```

   After you install the packages, you still have to connect the kernel to Jupyter. Type the following into R.

   ```
   > IRkernel::installspec()
   ```

   Type **q()** to quit from the kernel.

5. **Verify that the Kernel is Connected**

   Open a command line and type

   ```
   jupyter notebook
   ```

   A local server will open and a file list (see figure 2.2) will be displayed in your default web browser. On the top right is a tab `new`. Select `new` » `R notebook` from that tab. If `R notebook` is not available, something went wrong with the installation. If it is there, everything

---

[6]See https://irkernel.github.io/installation/ for further details.

was installed, and a new R notebook (see figure 2.3) should open in a new web browser window.

Figure 2.2.: The Jupyter file browswer opens when you start the Jupyter server by typing **jupyter notebook** at the command line. This will appear in your default web browswer. You can navigate to any sub-folder and select any file to load.

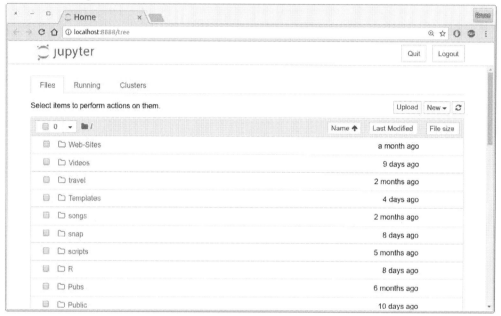

Opening Jupyter Notebooks

Jupyter notebooks have file extensions `.ipynb`.

The only way to open an existing file is by navigating to it through the Jupyter file browswer. You should never double click on the file to open it because this will not start the Jupyter server.

Rule Number 1

Never Ever Double Click on a Jupyter Notebook Icon To Open It!

Figure 2.3.: A new Jupyter notebook consists of a single blank cell, labled In[ ]:. As soon as you type something in this cell and press Enter, the contents of the cell will be evaluated, the label will become In[1]:, and a new blank cell labeled In[ ]: will appear beneat it. The icon at the top right tells you that this is an R notebook (and not a Python notebook).

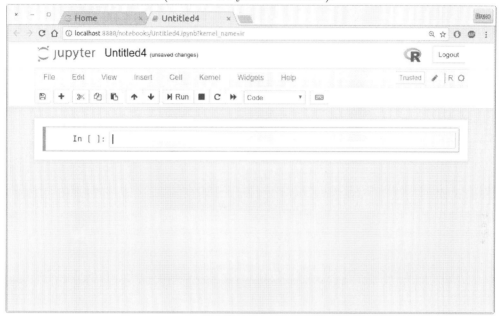

Rule Number 2

The only way to open an existing notebook is to navigate to that notebook through the Jupyter file browswer and select it there.

If you are coding in Windows you should probably enable **Show File Extensions** in the settings so that you can see which files are your notebook files. Windows will probably make their icons look like some sort of generic document, and will, depending on what you have installed on your system, tell you that you should open them in some bizarre and inappropriate application.

Similarly if you are working on a Mac (this is especially true when files are processed or have been downloaded through some learning management systems) the files will have an additional extension such as `.json` tagged onto the end of the file name. These files should be renamed to correct the file extension back to `.ipynb`.

3. Just Enough R

This is not a book on R, it is a book on machine learning. So don't expect to become a hotshot R programmer by reading this. Since all of the code examples are given in R, though, you should pick up a fair amount of R as you read through the book. This chapter will give you a brief introduction to R. If you are already familiar with R you can safely skip to the next chapter. For the most part, we will introduce new R capabilities (such as plotting) as we need them in the chapters on machine learning.

While no previous exposure to R is assumed, most readers will probably find it easier to have an R reference handy.[1] The R documentation website is a useful source if you know specifically what you are looking for, such as a particular function name,[2] and have good internet access, but it won't do you much good if your stuck on that long flight from Singapore to New York and the internet is down.[3] (Or if you live in my neighborhood.) All is not lost for those with spotty internet service, since a whopping big lot of manuals are available from CRAN including a 3500 page reference manual and a 105 page introduction to R.[4]

Keep in mind that unlike the vast majority of programming languages out there, R was not written for programmers, nor was it ever intended to be used as such. R was designed as an analytical tool, primarily intendend for statisticians. This does not mean that you can't write very complex programs in R. It just means that programming was not the central focus of its design. So if you are a hard-core programmer you will see that it violates practically every design paradigm that you have come to hold as holy writ. In fact, if you are a true programmer just learning R for the first time you should probably stop here and take a few deep breaths, because you will end up pulling out a lot of hair learning R.

What does this mean for the student? If you are completely new to programming, probably nothing. However, learning R will probably not help

---

[1] Such as Richard Cotton (2013) *Learning R*, O'Reilly. Beware that most books on R will try to teach you statistics at the same time.
[2] See https://www.rdocumentation.org/
[3] There is not a downloadable version.
[4] See https://cran.r-project.org/manuals.html.

prepare you for learning any other computer language in the future (with a few notable exceptions such as S, SAS, Mathematica). There are some vague similarities to the lispy languages, but not enough to make a difference. If you are comfortable programming in at least one procedural language (such as C, Java, or Python) be ready for a few surprises, try to keep an open mind, and don't go running for the hills.

About The Assignment Statement

In R the assignment statement normally uses a left-pointing arrow-like symbol. I say "arrow-like" because it requires the three key strokes, shift < - , to get something like **x<-5**. This means "(a) take the number 5; (b) store it in some machine location; (c) label that machine memory address with the 5 in it by **x**."

In the early days of computing other computer languages (e.g., APL, which had so many special symbols it required a special keyboard) toyed with assignment statements of this sort (or their equivalent, such as **x:=5** in ALGOL or Pascal) because mathematicians couldn't wrap their heads around statements of the form:

```
x = x+1
```

This is because mathematically the expression

$$x = x + 1$$

is equivalent to the statement

$$0 = 1$$

which makes no sense at all. So mathematicians and statisticans devised the left-pointing arrow to replace the simple equal sign for assignment.

Common sense prevailed in other languages, however. In languages like Python, Java or Fortran (and it is permitted in R as well, though it is not standard), we take **x=x+1** to mean (a) calculate the value of **x**; (b) add 1 to it; (c) store the result back in register **x**.

The standard that is used in all modern languages is that the equal sign is used for assignment.

Given that an equal sign requires one key stroke, and <- requres three keystrokes, we will in this book, keep with the of Ockham's Razer,[5] and always use the equal sign for assgnments in R.

> **Rule Number 3**
> Never use three keystrokes when one will do.

Using R

R can be used in several different ways:

1. Through the R console (calculator mode). Programs and modules can be written in a text editor or your favorite IDE and loaded and run, or typed in one line at a time. This method will not be discussed in this book.

2. Through a dashboard like RStudio. These dashboards divide the screen into windows for the console, editors, plots, help, etc. Some even provide rudimentary notebook support (not Jupyter compatible). This method will not be discussed in this book.

3. Through the Jupyter notebook interface. All examples shown in ths book will be illustrated using the Jupyter notebook.

To use the notebook, open a command line (**terminal** or **command prompt**) and type

```
jupyter notebook
```

The Jupiter notebook file navigator will open in your standard web broswer (see figure 2.2).

To open an existing notebook, use the file navigator menu to locate the

---

[5] Ockham's Razer is the philosophy that when faced with two (or more possible explanations), the simpler one is preferable to the more complex one. The engineering corollary to this is that when faced with two equivalent solutions, the simpler one is preferable.

notebook on your file system and click to open it.

To open a new notebook, click on the tab and the type right [new》R Notebook]. The notebook will be named **untitled**. To change the name select [File 》Rename ...].

Using R in Jupyter

A Jupyter notebook consists of a sequence of boxed areas called **Cells**. You need to place your code in cells.

Each cell has a cell type. You will see both input cells and output cells in your notebook. Input cells are labeled on the left by In [number] where **number** is some integer. Before a cell is evaluated, the number is blank. Each time any cell in the notebook is evaluated, an internal counter is incremented and this number is placed in the In label.

To evaluate a cell, place the cursor in a cell and type [shift] [enter]. (Hitting just [enter] alone will make the cell larger so you can place more lines of code in it.) You will usually have many input cells in a notebook and it is possible to evaluate them out of order (i.e., not from top to bottom). The numbers will tell you the order in which they have been evaluated.

Every cell in a Jupyter notebook has an output value. If this output value is not null, the results of the evaluation will be printed to the notebook immediately after the cell when you evaluate it. The output value will resemble the result of sending the result of the last statement in the cell to a **print** statement.

Once a cell has been evaluated, the results of all evaluations will be remembered for future evaluations, and will be known to code in other cells, even if this code is ordered earlier in the notebook. For this reason one should be very careful about evaluating cells out of order. You can reset all variables by reloading the kernel using as a menu option.

Input cells have different types as well. The main types you will use are **code** are **markdown**. To change the cell type place the cursor in the cell and select from the drop-down menu at the top of the page. Any text, html, latex, or markdown can be included in a markdown cell. This allows you to included documentation in-between blocks of code.

Packages and Libraries

So many extensions have been written for R that you will often find yourself needing to load an additional library. Libraries are in grouped into packages that you may or may not have already installed on your system. If a package is not already on your system you can usually get it either from CRAN or from another software archive.

To get a list of packages that can be intsalled, type

```
available.packages()
```

The list will be displayed as a data frame.

To download a package from CRAN (the online repository where most non-biology R packages are stored),

```
install.packages("packagename")
install.packages(c("package1", "package2","package3"))
```

Note that the package names must be enclosed in quotes.[6]

To use a library once it has been downloaded,

```
library(libraryname)
```

To install a package from a URL that is not CRAN, use

```
source("URL")}
```

where URL is replaced with the actual URL of the repository. For example, to install the biological tools in the bioconductor repository,

```
source("http://bioconductor.org/biocLite.R")
```

---

[6]Note that some linux systems are occasionally finicky about an occasional package and/or a dependency. If this happens to you, it may be necessary to install either the package itself or the dependency through your package manager.

```
Installing package into '/home/mathman/R/x86_64-pc-linux-gnu
-library/3.4' (as 'lib' is unspecified)
Bioconductor version 3.6 (BiocInstaller 1.28.0),
?biocLite for help
A new version of Bioconductor is available after installing
the most recent version of R;
see http://bioconductor.org/install
```

The recommending way of installing bioconductor is to use **biocLite**:

```
biocLite()
```

```
BioC_mirror: https://bioconductor.org
Using Bioconductor 3.6 (BiocInstaller 1.28.0), R 3.4.4
(2018-03-15). Installing package(s) 'Biobase', 'IRanges',
'AnnotationDbi' also installing the dependencies 'assertthat',
'bit', 'prettyunits', 'bit64', 'blob', 'pkgconfig', 'BH',
'plogr', 'BiocGenerics', 'S4Vectors', 'DBI', 'RSQLite'
```

To install a specific library from bioconductor[7] use **biocLite("library")**. The library name must be in quotes. Since all dependendencies will also be installed the process may take a while.

Identifers

Identifiers, or variable names, can consist of any combination of letters, numbers, the underscore character, and the period. An identifier must not begin with a digit, a period followed by a digit, or an underscore character.

An identifier cannot duplicate a reserved word (see table 3.1).

Table 3.1. Reserved Words in R.

break	if	NA_character_	NULL
else	in	NA_complex_	TRUE
FALSE	Inf	NA_integer_	while
for	next	NA_real_	repeat
function	NA	NaN	

---

[7]A current list of all libraries can be found at http://www.bioconductor.org/packages/release/BiocViews.html#___Software

Scalar Types in R

Scalar or **atomic** data types in R are `logical` (values are **TRUE** or **FALSE**), `integer` (32 bit representation), `double` (double precision floating point), `complex`, `character`, and `raw` (bytes).

There are conversion (casting) and testing functions for each data type. The testing functions are `is.integer(x)`, `is.double(x)`, ... The casting functions are `as.integer(x)`, `as.double(x)`, ...

Strings are represented by arrays of characters that are also called characters. Both `"a"` and `"spam and eggs"` have the `character` data type.

The function `typeof(x)` returns a string that gives the type of the identifier `x`.

Sequential Data Structures

A `vector` is a sequence of homoegenous items (items of the same data type). Vector elements are accesed by index (starting from 1). Vectors are stored in a contiguous area of memory and it is very inefficient to change the size of a vector after it has been initialized. Items can be assigned to a vector using the **combine** function **c**. Use the `length` function to determine the number of items in a vector:

```
x=c(1,2,3,4,5)
y=c("spam","and","eggs")
c(length(x), length(y))
```

    5 3

A sequence with a fixed interval is obtained with the `seq` functoin.

> `seq(from=`*value*`, to=`*value*`)`
> `seq(from=`*value*`, to=`*value*`, by=`*value*`)`
> `seq(from=`*value*`, to=`*value*`, length.out=` *value*`)`

The first form returns a sequence of numbers where the interval is 1. In the second the interval is explicitly stated. The last number of the vector returned is the greatest value in the sequence less than or equal to the `to=` value.

```
seq(from=5.0, to=15.0, by=1.5)
```

5 6.5 8 9.5 11 12.5 14

If the **length.out** form is used, the number of items is specified instead of the interval.

```
seq(from=5.0, to=18.5, length.out=10)
```

5 6.5 8 9.5 11 12.5 14 15.5 17 18.5

A shorthand for the first form is **from:to**.

```
5:10
```

5 6 7 8 9 10

When a vector is displayed as the return value of a cell, values may overlap onto multiple lines. In general, the indices are not displayed.

```
seq(1,9,length.out=27)
```

```
1 1.30769230769231 1.61538461538462 1.92307692307692
2.23076923076923 2.53846153846154 2.84615384615385
3.15384615384615 3.46153846153846 3.76923076923077
4.07692307692308 4.38461538461539 4.69230769230769
5 5.30769230769231 5.61538461538462 5.92307692307692
6.23076923076923 6.53846153846154 6.84615384615385
7.15384615384615 7.46153846153846 7.76923076923077
8.07692307692308 8.38461538461539 8.69230769230769 9
```

When a vector is displayed using the **print** command, it may take multiple lines. The index of the first item on each line is printed at the beginning of each line.

```
print(seq(1,9,length.out=27))
```

```
 [1] 1.000000 1.307692 1.615385 1.923077 2.230769 2.538462
 [7] 2.846154 3.153846 3.461538 3.769231 4.076923 4.384615
[13] 4.692308 5.000000 5.307692 5.615385 5.923077 6.230769
[19] 6.538462 6.846154 7.153846 7.461538 7.769231 8.076923
[25] 8.384615 8.692308 9.000000
```

You can change the line wrapping width with the **width** option as follows.

```
options(width=40) # line wrap after 40 characters
print(seq(1,9,length.out=27))
```

```
 [1] 1.000000 1.307692 1.615385 1.923077
 [5] 2.230769 2.538462 2.846154 3.153846
 [9] 3.461538 3.769231 4.076923 4.384615
[13] 4.692308 5.000000 5.307692 5.615385
[17] 5.923077 6.230769 6.538462 6.846154
[21] 7.153846 7.461538 7.769231 8.076923
[25] 8.384615 8.692308 9.000000
```

Lists are similar to vectors except that a list may include heterogeneous combinations of data types. Lists are created with the function **list**.

```
list(131, "cats")
```

1. 131
2. 'cats'

When displayed, the items in a list are numbered.

A slice (intermediate piece) of a vector between two indices can be extracted using the slice operator (the colon). For example, to extract the 13th through the 20th elements of the vector **stuff**,

```
stuff=5:25
stuff
```

5 6 7 8 9 10 11 12 13 14 15 16 17 18 19 20 21 22 23 24 25

```
stuff[13:20]
```

17 18 19 20 21 22 23 24

Slices can be extracted from the beginning of a vector **v** with **head(v)** and the end of **v** with **tail(v)**:

```
tail(stuff,5) # last 5 items
```

21 22 23 24 25

```
head(stuff,12) # first 12 items
```

```
 5  6  7  8  9 10 11 12 13 14 15 16
```
```
head(stuff,-5)  # everything but last 5 items
```
```
 5  6  7  8  9 10 11 12 13 14 15 16 17 18 19 20
```
```
tail(stuff,-5)  # everything but first 5 items
```
```
10 11 12 13 14 15 16 17 18 19 20 21 22 23 24 25
```

A negative index to **head** means count from the end, and a negative index to **tail** means count from the beginning.

Matrices and Arrays

An **array** is the basic data structure underlying both vectors and matrices in R. In fact both a **vector** and **matrix** are implemented as linear arrays. It is only when they are printed that the underlying structure becomes clear. Vectors and matrices are the building blocks of data frames.

A **vector** is an **array** with a single dimension, called its length.

A **matrix** a an **array** with two dimensions.

```
x=c(1,2,3,4,5)
y=c(2,4,6,8,10)
m=data.frame(x,y)
m
```

	x	y
1	1	2
2	2	4
3	3	6
4	4	8
5	5	10

The column headers are the same as the original vector names.

The rows can be accessed by index. **m[3]** will return

	x	y
3	3	6

Ch. 3. Just Enough R

`m[3,1]` will return the value 3, and `m[3,2]` will return the value 6.

`nrow(x)` returns the number of rows in an array `x`. In the example above, `nrow(m)` will return the number 5.

`ncol(x)` returns the number of columns in an array `x`. In the example above, `ncol(m)` will return the number 2.

```
x=c(1,2,3,4)
y=c(5,6,7,8)
z=c(9,10,11,12)
M=matrix(c(x,y,z), nrow=3)
print{M}
```

```
     [,1] [,2] [,3] [,4]
[1,]   1    4    7   10
[2,]   2    5    8   11
[3,]   3    6    9   12
```

In the Jupyter notebook, if you just type the name of a matrix, you get a formatted listing

```
M
```

1	4	7	10
2	5	8	11
3	6	9	12

Note that both arrays and matrices are column dominant, that is, the data fills up along the columns. When referring to elements, the first index in a two dimensional array gives the row, and the second index gives the column. To pull out a single row, include a comma after the row index:

```
M[2,]
```

2 5 8 11

To pull out a single column, include a comma before the column index:

```
M[,2]
```

4 5 6

To pull out a single element, specify both indices:

```
M[2,3]
```

```
8
```

In general an array can have as many dimensions as you like.

```
Z=array(c(x,y,z), dim=c(2,2,3))
Z
```

```
1 2 3 4 5 6 7 8 9 10 11 12
```

```
print(Z)
```

```
, , 1

     [,1] [,2]
[1,]    1    3
[2,]    2    4

, , 2

     [,1] [,2]
[1,]    5    7
[2,]    6    8

, , 3

     [,1] [,2]
[1,]    9   11
[2,]   10   12
```

A Bit About Printing

If you want to print a value in the middle of a calculation or during a loop you can't necessarily wait until all calculation finishes in a cell and the output of the cell just drops out on you when it is done. Furthermore you might want to print more than one thing. This is especially true when you are debugging code and you may want to see what is going on every other line. You need some kind of print for this.

The standard "print something out" statement in R is the **print** statement. Unfortunately, **print** is not as flexible as it is in most computer languages. This is because **print(x)** only lets you print a single R object. Furthermore, it will print a pesky **[1]** at the beginning of the line. This is useful for printing matrices and long vectors or lists but not for much else. The tabular printing in Jupyter notebooks pretty much makes the old **print** statement worthless. Look at this.

```
x=1
y=2.74237
z=3
v=c(x,y,z)
print(v)
```

[1] 1.00000 2.74237 3.00000

If want to put some annotation in front of it, like

```
print("The answer is",v)
```

[1] "The answer is"

it just ignores the second argument! You either have to put it on another line, getting another silly **[1]**,

```
print("The answer is")
print(v)
```

[1] "The answer is"
[1] 1.00000 2.74237 3.00000

or you can do some string processing. There are three common options used here:

1. The **paste** function. With **paste** you have to be careful because it will vectorize any text which could lead to unexpected results:

    ```
    paste("The answer is ", v)
    ```

    'The answer is 1'   'The answer is 2.74237'   'The answer is 3'

    One solution is to nest additional functions:

    ```
    paste("The answer is ",
          paste("(",toString(v), ")",
          sep=""))
    ```

```
'The answer is (1, 2.74237, 3)'
```

The **sep** option tells **paste** to join the items with a null space (the default is a blank space).

2. The **cat** function. Here **cat** is a string concatenation operator similar to **paste**.

```
cat("The answer is (", toString(v),")", sep="")
```

```
'The answer is  (1, 2.74237, 3)'
```

3. The **sprintf** function. It emulates the C function of the same name.

```
sprintf(
  "The answer is: (%-.1f, %5.3f, %3.1f) and |v| = %f",
  v[1],v[2],v[3],sqrt(v %*% v))
```

```
'The answer is: (1.0, 2.742, 3.0) and |v| = 4.185761'
```

Note that even in **sprintf**, there can only be a single string in the first argument. If you want to write it over multiple lines in your code, the line feed will automatically be replaced with a newline character in the output!

Of these, the third, **sprintf**, is the most flexible. It emulates the C **sprintf** function and similar encodings in other languages.

Data Frames

Most data analysis begins with data frames, rectangular arrays of numbers where each of the columns in labeled.

```
x=c(1,2,3,4)
y=c(5,6,7,8)
z=c(9,10,11,12)
p=data.frame(x,y,z)
p
```

x	y	z
1	5	9

2	6	10
3	7	11
4	8	12

The column names default to the names of the vectors from which the data frame is formed. They can also be reassigned at the time of creation.

```
p=data.frame("curly"=x,"moe"=y,"larry"=z)
```

The rows can be labeled with the function `row.names(frame)`.

```
row.names(p)=c("john","paul","george","ringo")
p
```

	curly	moe	larry
john	1	5	9
paul	2	6	10
george	3	7	11
ringo	4	8	12

There is also a function `colnames(frame)` that can be used to assign column names. To replace a single column name, use an index:

```
colnames(p)[2]="shem"
```

Loops

There are three basic types of loop structures in R: **for**, **repeat** and **while**.

The basic structure of a **for** loop is

> **for** (*name* **in** *vector*) *statement*

where *name* is any atomic identifier; *vector* is any vector identifier; and *statement* is any simple or block statement. The variable *name* will be sequentially assigned to each value in *vector*, and *statement* will be executed for that value before proceding to the next value in *vector*.

```
r=array(0,dim=c(10,3)) # 10 rows, 3 columns, zero fill
for (x in 1:10){
```

```
        xsquared = x*x
        xcubed = x*x*x
        xfourth = x*x*x*x
        r[x,]=c(xsquared, xcubed, xfourth)
    }
    r
```

```
  1     1     1
  4     8    16
  9    27    81
 16    64   256
 25   125   625
 36   216  1296
 49   343  2401
 64   512  4096
 81   729  6561
100  1000 10000
```

The basic structure of a **while** loop is

> **while (** *test* **)** *statement*

where *test* is any expression that evaluates to either **TRUE** or **FALSE**; and *statement* is any simple or block statement. The truth value of *test* is tested initially and at the start of each loop iteration. If the result is **TRUE**, then *statement* is executed. If the result of evaluating *test* is **FALSE** at the beginning of any loop iteration, then the loop is terminated.

```
s=array(0,dim=c(10,3))
x=1
while (x<=10){
    xsquared = x*x
    xcubed = x*x*x
    xfourth = x*x*x*x
    s[x,]=c(xsquared, xcubed, xfourth)
    x=x+1
}
```

The contents of the array **s** are identical to the contents of the array **r** produced in the example for the for **for** loop above.

The basic structure of a **repeat** loop is

repeat (*statement* **)**

where *statement* must be a block statement. The contents of *statement* are repeated indefinitely until a **break** is encountered.

```
q=array(0,dim=c(10,3))
x=1
repeat{
    xsquared = x*x
    xcubed = x*x*x
    xfourth = x*x*x*x
    q[x,]=c(xsquared, xcubed, xfourth)
    x=x+1
    if (x>10) break
}
```

Statements

The most common statement you will use in R is the assignment statement. In this book, we write an assignment statement as

    identifier = *expression*

An equivalent form (not recommended because it violates rule 3) is

    identifier <- *expression*

With the arrow notation, R allows you to put the target on the right hand side of the assignment. This is REALLY NOT RECOMMENDED because it is poor programming practice (and it also violates rule 3), unless you really want obfuscate your code.

    expression-> **identifier**

A Code cell in a Jupyter notebook can hold multiple statements. If these statements are written on different lines in the cell, no delimiter is required between the statements.

```
p=1
q=2
p+q
```

3

You can also string multiple statements along a single line in a code cell in a notebook so long as you end each statement by a semicolon.

```
p=1; q=2; p+q
```

3

Conditional statements, loops, and function definitions may include blocks of code. A **block** of code is a sequence of statements enclosed in curly brackets. For clarity it is recommended that you always indent your code when using a block. There are exceptions, such as in short function definitions. However, not indenting your code will lead to code obfuscation.

> **Rule Number 4**
> Code blocks should usually be indented.

Conditional Control in R

Conditional evaluation in R is performed by `if` and `switch`. Technically, `switch` is a function while `if` is a statement.

The `if` statement has the form:

> `if` (*condition*) *statement*

where *condition* is an expression that evaluates to **TRUE** or **FALSE** and *statement* is either a simple statement or a block. If the value of *condition* after evaluation is **TRUE**, then the *statement* is executed. Otherwise, it is not exectuted.

The `if` statement with `else` has the format

> `if` (*condition*)
>   *statement1*
> `else`

statement2

where *statement1* and *statement2* are either simple statements or blocks. (The indentation illustrated here is optional.) If *test* evaluates to **TRUE** than statement1 is evaluated and if *test* evaluates to **FALSE** then statement2 is evaluated. The **if** and **else** statements can be nested:

> **if** (*condition1*)
>   *statement1*
> **else if** (*condition2*)
>   *statement2*
> **else if** (*condition3*)
>   *statement3*
> **else**
>   *statement4*

where each condition evaluates to either **True** or **False**, and each statement may either be an atomic statement or a block.

In the following example of a nested **if** we use the function **sapply(***vector, function***)**, which applies the function to each element of a function.

```
decade=function(x){
    if (x<10)
        0
    else if (x<20)
        1
    else if (x<30)
        2
    else 3
}
sapply(seq(5,35,10), decade)
```

0 1 2 3

Defining Functions

Function definitions have the following forms form:

> **function**(*arglist*) *expression*
> **function**(*arglist*){*expression list*}

$$functionname = \textbf{function}(arglist)\{expression\ list\}$$

The return value of the function is the value of the last expression evaluated. The curly braces are optional if only a single expression is included. If a name is assigned to the function, the function can be invoked later by name:

```
my2square = function(x) {z=x*x; 2*z}
my2square{17}
```

```
578
```

If the function is only needed once, it can be used as a lambda expression (without assigning a name) embedded in a larger expression.

Plotting

To plot points $(x_1, y_1), \ldots, (x_1, y_n)$ in vectors **x** and **y**:

```
plot(x,y)
```

Optional parameters (table 3.2) can be specified in the function call by calling the function **par**.

Table 3.2. Optional Parameters that can be use with **plot**. These must be specified as **name=value** in the invocation.

Parameter	Description
main	Plot title.
sub	Subtitle.
xlab	x axis label.
ylab	y axis label.
asp	Aspect ratio.

To add a line $y = a + bx$, or the line *linearmodel* that is produced by **lm**, to a plot:

```
abline(a,b)
abline(linearmodel)
```

Ch. 3. Just Enough R

To add a smoothe curve connecting points $(x_1, y_1), \ldots, (x_1, y_n)$ in vectors **x** and **y**:

```
lines(x,y)
```

Optional parameters (such as line style) can be specified in the function call by calling the function **par** (see table 3.3).

Table 3.3. A few of the optional graphics parameters can be specified as **name=value** pairs in a call to **par** or in a call to a plotting function.

Parameter	Description
ann	**TRUE** (default) to enable or disble all annotations.
box	Determines which edges of the plot are drawn. Values of options resemble shape of resulting box. Can also use **box** function. **"o"** (default; draw entire box) **"c"** (omit right edge) **"u"** (omit top edge) **"7"** (omit left and bottom) **"j"** (omit left and top edges) **"l"** (only left edge) **"n"** (no box)
cex	Multiply size of points to be plotted. Default is 1.
col	Color value. The function **colors** returns a list of allowed values. Also available: **col.axis** Axis color **col.lab** Label color **col.main** Main title color **col.sub** Subtitle color **bg** Background color **fg** Foreground color
family	Font family, such as **"serif"**, **"sans"**, or **"mono"**.
font	Font style: 1 (plain), 2 (bold), 3 (italic), 4 (bold italic). Also available: **font.axis** **font.lab** **font.main** **font.sub**
lty	Line type. **"blank"**, **"solid"**, **"dashed"**, **"dotted"**, **"dotdash"**, **"longdash"**, **"twodash"**
lwd	Line width. Default is 1.
pch	Symbol to use to plot points. Typical values: 19 Solid Circle 23 Solid Diamond 20 Small Solid Circle 24 Triangle Pointing Up 21 Filled circle 25 Triangle Pointing Down 22 Filled square

Sampling

```
sample(x, size)
sample(x, size, replace = FALSE, prob = NULL)
```

If **replace** is not **FALSE** then sampling is performed with replacement. The same item may be selected multiple times:

```
x=c("tom","dick", "harry","curly","moe","larry")
sample(x,5,TRUE)
```

'dick' 'dick' 'harry' 'curly' 'dick'

If **replace** is not specified, the sampling is done without replacement, and there will be no duplicates.

```
x=c("tom","dick", "harry","curly","moe","larry")
sample(x,5,TRUE)
```

'dick' 'harry'

The actual output depends upon the random seed set in the machine. You can change this:

```
set.seed(number)
```

where **number** is replaced by an integer value.

The **prob** argument to **sample** is an optional vector of probabilities for sampling. If it is not specified, all elements in the array have the same probability of being sampled.

Part II.
Regression

Regression is the process of finding relationships between different variables. Usually this means finding curves that best represent points on a scatterplot. There are two broad classes of regression we will consider:

- **Linear Regression** - The relationship is linearly dependent on a collection of unknown parameters. We can fit lines, planes, and polynomials. We can solve for the coefficients analytically (although the formula may be very messy or involve large matrices, and so we will still need a computer to actually find them).
- **Nonlinear Regression** - The relationship depends non-linearly on the unknown parameters. Usually these methods require an iterative process such as gradient descent and the solution may not be unique. These methods include most curve fitting methods as well as neural networks.

4. Linear Regression

The idea behind linear regression is this: given a scatter plot of (x_i, y_i) data pairs (see fig. 4.1), what is the equation of the line $\hat{y} = a + bx$ that best describes the data? Here we write a predicted value $\hat{y}$ to distinguish it from an observed value y. The scatter plot of all the (x, y) pairs will generally form a cloud of data. The "best" line will not, in general, pass through any of the points, but will pass through the center of the cloud of pairs, where we are using the word "center" in a very non-specific way.

Figure 4.1.: A scatter plot of a linear data set usually consists of a cloud of data. The best fit line will, in some sense, pass through the mid-line of this cloud.

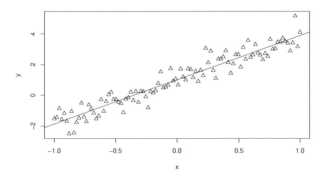

The mathematics behind this is described at the end of the chapter. Briefly, the best fit line is one that minimizes the sum of the squares of the vertical distances of all the observations. Formulas for the slope and and intercept can be calculated exactly using calculus. This method can be extended to higher dimensions (planes and hyperplanes) and polynomials as discussed in chapters 5 and 6.

Linear Regression in R

To illustrate regression we will use the two dimensional data set **avtemp.csv**. This file contains two columns of data. Each row has two values, separated by commas. The first row contains the text labels, also separated by commas. The data represents global average temperature from 1800 through

2017, one point per year, as degrees above or below the average from 1951 through 1980.

We read the file into a data frame with the R function **read.table**. The function **colnames** can be used to give the names of the columns, which are determined from the first row of the data set.

```
temperatures=read.table("avtemp.csv", TRUE, sep=",")
colnames(temperatures)
```

'year' 'tempdiff'

If we ask R to list the data in a Jupyter notebook, it gives a nice, formatted listing (be sure to select the cell 〉 current output 〉 toggle scrolling if you are listing a long table.

temperatures

year	tempdiff
1880	-0.19
1881	-0.10
1882	-0.10
1883	-0.19
1884	-0.28
1885	-0.31
..	..
2015	0.86
2016	0.99
2017	0.90

We can verify that the average from 1951 to 1980 is very close to zero. We select the subset of temperatures of between these years and add them up. **subset** selects the desired rows, and index **[2]** picks the second column, after we have selected the desired rows. (The first column is the year, second column has temperature differences.) Then we add them up and divide by the number of rows.

```
sum(subset(temperatures,year>1950 & year <= 1980)[2])/
   (1980-1950)
```

2.31296463463574e-19

We can get the data from individual columns with the column select operator (the dollar sign):

---

34   Ch. 4. Linear Regression

```
years=temperatures$year
temps=temperatures$tempdiff
plot(years,temps, xlab="Year of Data Set (1880-1917)",
     ylab="Excess Temperature (over 1951-1980 avg.) Deg. C",
     main="Global Average Temperature")
```

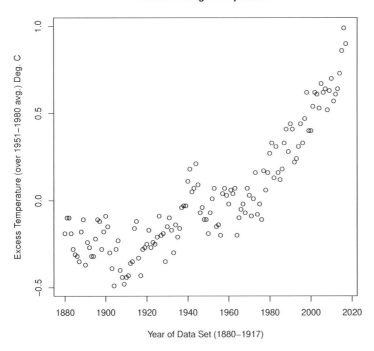

The plot is normally displayed inline in your Jupyter notebook, directly following the **plot** command. To save the plot to a pdf file, use

```
pdf(my_filename.pdf)
plot(years, temps, ... )
dev.off()
```

For other graphics formats use instead of **pdf** use **postscript**, **svg**, **png**, **jpeg**, **bmp**, **tiff**, etc.

Lets try to fit a straight line to this data. This process is called linear regression. To be able to test the accuracy of of our fit, rather than fitting all the data, we will split up our data into two parts:

- Training data - the training data is used to fit the model coeficients a and b in the line $y = a + bx$.

- Test data - the test data is used to test how good the fit is. We use the equation $y = a + bx$ determined from the training data but use x values from the test set. This gives us predicted y values that we can compare with the y values in the test data set.

We can split up the original data into test and training sets any way we want. A typical split is 75% training/25% test data.

Figure 4.2.: Separation of data data into test and training sets.

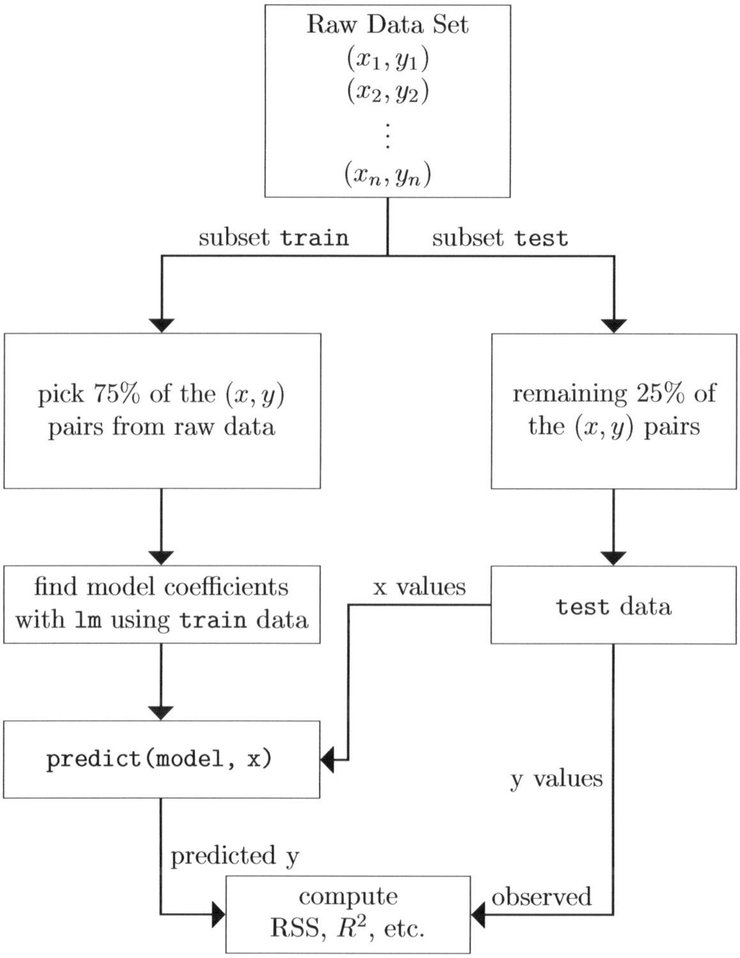

Here is one way very direct way to split the data up. If there are n rows of data, generate a sequence of integers $1, 2, 3, ..., n$; you can do this in R by simply writing the **1:n**. The function **sample** will randomly pick a subset of these. To get around 75 percent,

```
n = nrow(temperatures)
index = sample(1:n,round(0.75*n))
```

The variable **index** then contains a vector of around 75% of the numbers $1, 2, 3, ..., n$ selected randomly. We can select just those rows that are specified by **index** using **temperatures[index,]**, and we can select the remaining rows using **temperatures[-index,]**. Use this to obtain our test and training sets.

```
train=temperatures[index,]
test=temperatures[-index,]
```

We perform the least squares fit using the linear model fit function **lm**. The general syntax of the **lm** function is

```
lm(response~explanatory,options,)
lm(response~var1+var2+var3,options,)
lm(response~.,options,)
lm(response~.-skip,options,)
```

where *options* are any options (table 4.1; and **response**, **explanatory**, **vari**, **skip** are the column names (variables) of any variables in the data frame. The first form solves for an equation of the form

$$(\text{\textbf{response}}) = a + b(\text{\textbf{explanatory}})$$

The second form solves for a multilinear response of the form

$$(\text{\textbf{response}}) = a + b_1(\text{\textbf{var1}}) + b_2(\text{\textbf{var2}}) + \cdots$$

There may be fewer or more variables. The fourth form solve for a response of the form

$$(\text{\textbf{response}}) = a + \sum_{\text{all columns}} b_i \text{\textbf{var}}_i$$

where the sum is over all variables except the response variable. The fourth versions is the same, except that it also omits the variable **skip**.

Table 4.1. Parameters to `lm`

Parameter	Description
`contrasts`	Optional contrasts matrix. See https://www.rdocumentation.org/packages/stats/versions/3.5.0/topics/model.matrix
`data`	Data frame or matrix containing the data to be fit.
`formula`	A formula such as `y~x`
`na.action`	What to do when data is missing. Default is `na.omit` (remove this entire line of data); alternatives are `na.fail`; `na.exclude` (pad the data).
`subset`	Optional vector specifying a subset of the data to be used.
`weights`	Optional numeric weight vector. If weights are given then in stead of a standard least squares fit, each observation is weighted, the weight is inversely proportional to the variance.

Return Values	
`call`	Call to the function.
`coefficents`	vector of coefficients.
`contrasts`	Contrasts used (where requested).
`df.residual`	Degrees of freedom.
`fitted.values`	Fitted mean values.
`model`	Resulting model.
`na.action`	How NA's were handled.
`offset`	Offset values.
`rank`	Rank of model.
`residuals`	Residuals to fitted values (observed minus expected).
`terms`	Terms in the formula
`weights`	Weights found (where requested).
`x`	Table of x values.
`xlevels`	Record of the levels used (where requested)
`y`	Table of y values.

Here we save the result in the variable `lm.fit`.

```
lm.fit = lm(tempdiff~year, data=train)
```

The first argument `tempdiff~year` is a formula. It says that the linear model should fit an equation $y = a + bx$ using x data taken from the column labeled `year` in `train` and y data in the column labeled `tempdiff`, i.e., fit the model (`tempdiff`) $= a + b($`year`$)$. The ultimate output of the linear model are the numbers a and b; however, we can get a whole lot more information from `summary`. For detailed parameters for `lm` see table 4.1.

```
print(summary(lm.fit))
```

```
Call:
lm(formula = tempdiff ~ year, data = train)

Residuals:
     Min       1Q   Median       3Q      Max
-0.34286 -0.12176 -0.00131  0.13167  0.47341

Coefficients:
              Estimate Std. Error t value Pr(>|t|)
(Intercept) -1.397e+01  8.539e-01  -16.36   <2e-16 ***
year         7.187e-03  4.379e-04   16.41   <2e-16 ***
---
Signif. codes:  0 '***' 0.001 '**' 0.01 '*' 0.05 '.' 0.1 ' ' 1

Residual standard error: 0.1723 on 102 degrees of freedom
Multiple R-squared:  0.7254,    Adjusted R-squared:  0.7227
F-statistic: 269.4 on 1 and 102 DF,  p-value: < 2.2e-16
```

We can use the function **abline** to add the straight line predicted by the fit to our plot.

```
plot(years,temps, xlab="Year of Data Set (1880-1917)",
    ylab="Excess Temperature (over 1951-1980 avg.) Deg. C",
    main="Global Average Temperature and Least Squares Fit")
abline(lm.fit)
```

The resulting plot is shown in figure 4.3.

That doesn't really look like a great fit. One way to measure the accuracy of the prediction using the *test* data is to caclulate the *residual sum square error*. The residual error is the difference between the predicted y value, denoted by $\hat{y}$ (where $\hat{y} = a + bx$), and the observed y value, at a fixed value of x. The square of the resisual is $(\hat{y} - y)^2$. If we sum the squares of the residual errors over all possible (x, y) pairs, we get the the RSS error:[1]

$$\text{RSS} = \frac{1}{n} \sum_{\substack{\text{test} \\ \text{data}}} (y_i - \hat{y}_i)^2 = \frac{1}{n} \sum_{\substack{\text{test} \\ \text{data}}} (y_i - a - bx_i)^2$$

---

[1] Some authors may call this the *average* or *mean* residual sum square error since we are dividing by n.

Figure 4.3.: Plot of original data and least squares fit with `lm` using `abline`.

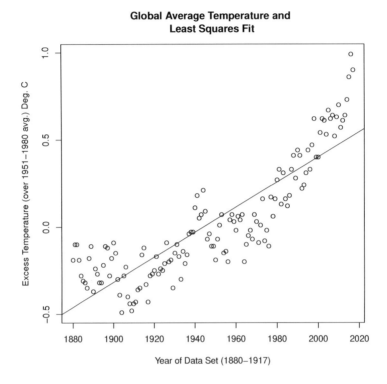

Here n is the number of data points in test data; a and b are the intercept and slope predicted by the linear model; and $(x_1, y_1), \ldots, (x_n, y_n)$ is the test data. None of the training data is used to calculate the RSS.

Another measure that can be used is the R^2 value

$$R^2 = 1 - \frac{\text{RSS}}{\text{MSS}}$$

where MSS, the *Mean Sum Squares* error, is the Average of the squares of the differences of the observed y_i values from their means.

$$\text{MSS} = \frac{1}{n} \sum_{\substack{\text{test} \\ \text{data}}} (y_i - \overline{y})^2$$

Here $\bar{y}$ is the mean of the y_i values,

$$\bar{y} = \frac{1}{n} \sum_{\substack{\text{test} \\ \text{data}}} y_i$$

To calculate the RSS and R squared, we begin with the function **predict(** *model, dataframe***)**. This returns a vector of y-values predicted by the model ($a + bx$ values) based on x data in the dataframe. There must be a column in the input *dataframe* argument that has the same name as the x value in the model.

```
ypredicted=predict(lm.fit,test)
```

This works because **lm.fit** was calculated using a model **tempdiff~year** and the dataframe **test** has a column labeled **year** in it. So **predict** uses the data in the **year** column to produce a vector of y predictions.

We can extract the observed value as a vector,

Since R does vectorized calculations, when we subtract one vector from another, the result is a new vector containing all the differences. So the difference between the vectors **ypredicted-yobserved** is a vector that gives the differences $a - bx_i - y_i$. If square this vector, using **(ypredicted-yobserved)^2** we get a vector where the ith element is $(a - bx_i - y_i)^2$. We want the sum of these, which we can calculate with the **sum** function.

```
yobserved=test$tempdiff    # $ selects a column
ypredicted=predict(lm.fit,test)
sum((ypredicted-yobserved)^2)/nrow(test)
```

0.0207858530770105

This is a number we can use to compare with other fits.

How Linear Regression Works[2]

In linear regression, the target function is $f(x) = a + bx$ for some undetermined constants a and b. We can define an **energy function** or **objective function** $\mathcal{E}(a,b)$ to be minimized by

$$\mathcal{E}(a,b) = \frac{1}{2}\sum_{i=1}^{n}|f(x_i) - y_i|^2 = \sum_{i=1}^{n}(a + bx_i - y_i)^2$$

The unknown constants a and b are determined by equating the partial derivatives $\partial\mathcal{E}/\partial a$ and $\partial\mathcal{E}/\partial b$ to zero. Energy is a concept that is stolen from physics and widely used in machine learning. It is a any function that we whose minimum corresponds to our objective or goal. The factor of $1/2$ does not affect the result because of the linearity of the derivatives and is placed there for convenience.

Figure 4.4.: The sum of all the vertical distances is minimized in linear regression.

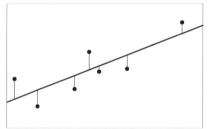

$$0 = \frac{\partial \mathcal{E}}{\partial a} = \frac{\partial}{\partial a}\frac{1}{2}\sum_{i=1}^{n}(a + bx_i - y_i)^2 = \sum_{i=1}^{n}(a + bx_i - y_i)$$
$$= \sum_{i=1}^{n} a + \sum_{i=1}^{n} bx_i + -\sum_{i=1}^{n} y_i = na + b\sum_{i=1}^{n} x_i - \sum_{i=1}^{n} y_i$$
$$= na + bX - Y$$

where

$$X = \sum_{i=1}^{n} x_i \text{ and } Y = \sum_{i=1}^{n} y_i$$

---

[2]The mathematical content of this and the following sections are based on the author's *Scientific Computation (3rd ed.)*, Ch. 37, "Least Squares."

Similarly,

$$
\begin{aligned}
0 &= \frac{\partial \mathcal{E}}{\partial b} = \frac{\partial}{\partial b} \frac{1}{2} \sum_{i=1}^{n} (a + bx_i - y_i)^2 = \sum_{i=1}^{n} x_i (a + bx_i - y_i) \\
&= a \sum_{i=1}^{n} x_i + \sum_{i=1}^{n} bx_i^2 - \sum_{i=1}^{n} x_i y_i = aX + b \sum_{i=1}^{n} x_i^2 - \sum_{i=1}^{n} x_i y_i \\
&= aX + bA - C
\end{aligned}
$$

where

$$A = \sum_{i=1}^{n} x_i^2 \text{ and } C = \sum_{i=1}^{n} x_i y_i$$

This gives us a a system of two linear equations in two unknowns a and b:

$$0 = na + bX - Y$$
$$0 = aX + bA - C$$

Multiplying the first equation by A and the second by X gives

$$
\begin{aligned}
0 &= A(na + bX - Y) = Ana + AXb - AY \\
0 &= X(aX + bA - C) = aX^2 + AXb - CX
\end{aligned}
$$

Subtracting,

$$0 = Ana - AY - aX^2 + CX = a(An - X^2) + CX - AY$$

and therefore

$$a = \frac{AY - CX}{An - X^2} = \frac{\sum_{i=1}^{n} x_i^2 \sum_{i=1}^{n} y_i - \sum_{i=1}^{n} x_i y_i \sum_{i=1}^{n} x_i}{n \sum_{i=1}^{n} x_i^2 - \left(\sum_{i=1}^{n} x_i\right)^2}$$

Alternatively, we could have multiplied the first equation by X and the second equation by n:

$$0 = X(na + bX - Y) = naX + bX^2 - YX$$
$$0 = n(aX + bA - C) = naX + nAb - nC$$

Subtracting,
$$0 = b\left(X^2 - nA\right) - (YX - nC)$$
Solving for b gives Y, gives
$$b = \frac{XY - nC}{X^2 - nA} = \frac{\sum_{i=1}^{n} x_i \sum_{i=1}^{n} y_i - n \sum_{i=1}^{n} x_i y_i}{\left(\sum_{i=1}^{n} x_i\right)^2 - n \sum_{i=1}^{n} x_i^2}$$

Derivation in Matrix Form

Instead of solving the equations by substitution we could have used matrix algebra. In matrix form,
$$an + b \sum_{i=1}^{n} x_i = \sum_{i=1}^{n} y_i$$
$$a \sum_{i=1}^{n} x_i + b \sum_{i=1}^{n} x_i^2 = \sum_{i=1}^{n} x_i y_i$$
In matrix form, this becomes
$$\begin{bmatrix} n & \sum x_i \\ \sum x_i & \sum x_i^2 \end{bmatrix} \begin{bmatrix} a & b \end{bmatrix} = \begin{bmatrix} \sum y_i \\ \sum x_i y_i \end{bmatrix}$$
Solving the matrix equation for $[a, b]^{\mathrm{T}}$ gives the same result.

Statistical Form

It is sometimes useful to express the quantities A, C, D, X, and Y in terms of known statistical quantities, in particular, the sample standard deviations s_x and s_y, where
$$s_x^2 = \frac{\sum_i (x_i - \bar{x})^2}{n - 1}, \text{ and } s_y^2 = \frac{\sum_i (y_i - \bar{y})^2}{n - 1},$$
where the sample means are $\bar{x} = \dfrac{1}{n}\sum x_i = X/n$ and $\bar{y} = \dfrac{1}{n}\sum y_i = Y/n$; and the correlation
$$r = \frac{1}{n-1} \sum_i \left[\frac{x_i - \bar{x}}{s_x}\right]\left[\frac{y_i - \bar{y}}{s_y}\right].$$

Since s_x and s_y are constants in this sum,

$$\begin{aligned}(n-1)rs_xs_y &= \sum_i (x_i - \bar{x})(y_i - \bar{y}) \\ &= \sum_i (x_i y_i - \bar{x} y_i - \bar{y} x_i + \overline{xy}) \\ &= \sum_i x_i y_i - n\overline{xy} = C - \frac{1}{n}XY\end{aligned}$$

Similarly, the observed sample variance s_x^2 satisfies

$$\begin{aligned}(n-1)s_x^2 &= \sum_i (x_i - \bar{x})^2 \\ &= \sum (x_i^2 - 2x_i\bar{x} + \bar{x}^2) \\ &= \sum x_i^2 - n\bar{x}^2 = A - \frac{1}{n}X^2\end{aligned}$$

Therefore

$$m = \frac{XY - nC}{X^2 - nA} = \frac{C - \frac{1}{n}XY}{A - \frac{1}{n}X^2} = \frac{(n-1)rs_xs_y}{(n-1)s_x^2} = r\frac{s_y}{s_x}$$

Similarly,

$$\begin{aligned}b &= \frac{AY - CX}{An - X^2} = \frac{AY - CX}{n(n-1)s_x^2} \\ &= \frac{AY - X(\frac{1}{n}XY + (n-1)rs_xs_y)}{n(n-1)s_x^2} \\ &= \frac{AY - X(\frac{1}{n}XY)}{n(n-1)s_x^2} - \frac{X((n-1)rs_xs_y)}{n(n-1)s_x^2} \\ &= \frac{Y}{n} \times \frac{(A - X^2/n)}{(n-1)s_x^2} - \frac{X}{n} \times \frac{rs_y}{s_x} \\ &= \frac{1}{n}Y - \frac{1}{n}mX = \bar{y} - m\bar{x}\end{aligned}$$

5. Polynomial Least Squares Regression

The term *polynomial regression* is something of a misnomer. What we really should say is *least squares linear regression of a polynomial*. What we are really doing here is finding the **best fit polynomial of degree** n to a data set. We solve for coefficients using a least squares process. Because the resulting equations *linear* in the unknown coefficients, the regression is still linear, just the result is nonlinear.[1]

Mathematical Background[2]

We can generalize the least squares problem to a polynomial of any degree n. Suppose we replace the objective function used previously with a polynomial of degree $n-1$

$$P(x) = c_1 + c_2 x + \cdots + c_n x^{n-1}$$

If there are m data points $(x_1, y_1), \ldots, (x_m, y_m)$, then a perfect fit would give

$$c_1 + c_2 x_1 + \cdots + c_n x_n^{n-1} = y_1$$
$$\vdots$$
$$c_1 + c_2 x_m + \cdots + c_n x_m^{n-1} = y_m$$

However, in any practical problem $m > n$ and usually $m \gg n$, so the system is (extremely) overdetermined. Thus

$$\begin{bmatrix} 1 & x_1 & x_1^2 & \cdots & x_1^{n-1} \\ 1 & x_2 & x_2^2 & & x_2^{n-1} \\ \vdots & & & & \\ 1 & x_m & x_m^2 & \cdots & x_m^{n-1} \end{bmatrix} \begin{bmatrix} c_1 \\ c_2 \\ \vdots \\ c_m \end{bmatrix} = \begin{bmatrix} y_1 \\ y_2 \\ \vdots \\ y_m \end{bmatrix} + \mathbf{r}$$

where $\mathbf{r}$ is a vector of **residuals** to be minimized. Our goal is find the vector $\mathbf{c}$ that minimizes $\mathbf{r}$, where

$$\mathbf{Ac} = \mathbf{y} + \mathbf{r}$$

---

[1] Many authors confuse the two concepts.
[2] The advanced mathematics in this section can be skipped without loss of continuity.

i.e., find **c** to minimize the distance

$$|\mathbf{r}| = |\mathbf{Ac} - \mathbf{y}|$$

If we denote the jth column of **A** by $\mathbf{a}_j$, where

$$\mathbf{a}_j = \begin{bmatrix} x_1^{j-1} \\ x_2^{j-1} \\ \vdots \\ x_m^{j-1} \end{bmatrix}$$

then

$$\mathbf{A} = \begin{bmatrix} \mathbf{a}_1 \mid \mathbf{a}_2 \mid \cdots \mid \mathbf{a}_{n-1} \end{bmatrix}$$

and

$$\mathbf{Ac} = c_1 \mathbf{a}_1 + c_2 \mathbf{a}_2 + \cdots + c_n \mathbf{a}_n = \sum c_j \mathbf{a}_j$$

We minimize the objective function

$$\mathcal{E}(c_1, \ldots, c_n) = |\mathbf{r}|^2 = \mathbf{r}^T \mathbf{r} = (\mathbf{Ac} - \mathbf{y})^T (\mathbf{Ac} - \mathbf{y})$$

by setting the derivatives $\partial \mathcal{E} / \partial c_i = 0$:

$$\begin{aligned} 0 &= \frac{\partial}{\partial c_i} \left((\mathbf{Ac} - \mathbf{y})^T (\mathbf{Ac} - \mathbf{y}) \right) \\ &= \left(\frac{\partial}{\partial c_i} (\mathbf{Ac} - \mathbf{y})^T \right) (\mathbf{Ac} - \mathbf{y}) + (\mathbf{Ac} - \mathbf{y})^T \frac{\partial}{\partial c_i} (\mathbf{Ac} - \mathbf{y}) \end{aligned}$$

But from (5)

$$\frac{\partial}{\partial c_i} \mathbf{Ac} = \mathbf{a}_i$$

Thus

$$0 = (\mathbf{a}_i)^T (\mathbf{Ac} - \mathbf{y}) + (\mathbf{Ac} - \mathbf{y})^T \mathbf{a}_i$$

Distributing the transposes,

$$0 = \mathbf{a}_i^T \mathbf{Ac} - \mathbf{a}_i^T \mathbf{y} + \mathbf{c}^T \mathbf{A}^T \mathbf{a}_i - \mathbf{y}^T \mathbf{a}_i$$

Rearranging,

$$\mathbf{a}_i^T \mathbf{y} + (\mathbf{a}_i^T \mathbf{y})^T = \mathbf{a}_i^T \mathbf{Ac} + (\mathbf{a}_i^T \mathbf{Ac})^T$$

Each of these terms is a number, so we can omit the transpose on the second term on each side of the equation,

$$2\mathbf{a}_i^\text{T}\mathbf{y} = 2\mathbf{a}_i^\text{T}\mathbf{A}\mathbf{c}$$

Dividing by 2 and taking into account that the equation is true for all $i = 1, 2, \ldots, n-1$,

$$\mathbf{A}^\text{T}\mathbf{y} = \mathbf{A}^\text{T}\mathbf{A}\mathbf{c} \qquad \text{\textbf{Normal Equations}}$$

This last equation is called the **Normal Equations** of the system. The normal equations can be simplified to

$$\mathbf{M}\mathbf{c} = \mathbf{b}$$

where $\mathbf{M} = \mathbf{A}^\text{T}\mathbf{A}$ and $\mathbf{b} = \mathbf{A}^\text{T}\mathbf{y}$. The elements of the matrix $\mathbf{M}$ and the vector $\mathbf{b}$ *are completely determined by the input data.* The only unknowns are the coefficients of the polynomial, which are listed from lowest order to highest order, in the vector $\mathbf{c}$. The matrix $\mathbf{M} = \mathbf{A}^\text{T}\mathbf{A}$ is an $n \times n$ symmetric matrix, known as the **normal matrix** given by

$$\mathbf{A}^\text{T}\mathbf{A} = \begin{bmatrix} 1 & 1 & \cdots & 1 \\ x_1 & x_2 & \cdots & x_m \\ x_1^2 & x_2^2 & \cdots & x_m^2 \\ \vdots & & & \vdots \\ x_1^{n-1} & x_2^{n-1} & \cdots & x_m^{n-1} \end{bmatrix} \begin{bmatrix} 1 & x_1 & x_1^2 & \cdots & x_1^{n-1} \\ 1 & x_2 & x_2^2 & & x_2^{n-1} \\ \vdots & & & & \\ 1 & x_m & x_m^2 & \cdots & x_m^{n-1} \end{bmatrix}$$

$$= \begin{bmatrix} m & \sum x_i & \sum x_i^2 & \sum x_i^3 & \cdots & \sum x_i^{n-1} \\ \sum x_i & \sum x_i^2 & \sum x_i^3 & \cdots & & \vdots \\ \sum x_i^2 & \sum x_i^3 & & & & \\ \sum x_i^3 & & & & & \\ \vdots & & & & & \vdots \\ \sum x_i^{n-1} & \cdots & & & \cdots & \sum x_i^{2(n-1)} \end{bmatrix}$$

The right hand side of the simplified form of the normal equations is given by the column vector

$$\mathbf{b} = \mathbf{A}^\text{T}\mathbf{y} = \begin{bmatrix} 1 & 1 & \cdots & 1 \\ x_1 & x_2 & \cdots & x_m \\ x_1^2 & x_2^2 & \cdots & x_m^2 \\ \vdots & & & \vdots \\ x_1^{n-1} & x_2^{n-1} & \cdots & x_m^{n-1} \end{bmatrix} \begin{bmatrix} y_1 \\ y_2 \\ \vdots \\ y_m \end{bmatrix} = \begin{bmatrix} \sum y_i \\ \sum x_i y_i \\ \sum x_i^2 y_i \\ \vdots \\ \sum x_i^{n-1} y_i \end{bmatrix}$$

Thus the normal equations in their full glory become

$$\begin{bmatrix} m & \sum x_i & \sum x_i^2 & \sum x_i^3 & \cdots & \sum x_i^{n-1} \\ \sum x_i & \sum x_i^2 & \sum x_i^3 & \cdots & & \vdots \\ \sum x_i^2 & \sum x_i^3 & & & & \\ \sum x_i^3 & & & & & \\ \vdots & & & & & \vdots \\ \sum x_i^{n-1} & \cdots & & & \cdots & \sum x_i^{2(n-1)} \end{bmatrix} \begin{bmatrix} c_1 \\ c_2 \\ \vdots \\ c_{n-1} \end{bmatrix} = \begin{bmatrix} \sum y_i \\ \sum x_i y_i \\ \sum x_i^2 y_i \\ \vdots \\ \sum x_i^{n-1} y_i \end{bmatrix}$$

If $n = 2$, so that we are fitting a polynomial of degree $n - 1 = 2 - 1 = 1$ (a line), the normal equations reduce to:

$$\begin{bmatrix} m & \sum x_i \\ \sum x_i & \sum x_i^2 \end{bmatrix} \begin{bmatrix} c_1 \\ c_2 \end{bmatrix} = \begin{bmatrix} \sum y_i \\ \sum x_i y_i \end{bmatrix}$$

These are the same equations we had for linear regression, with c_1 being the y intercept and c_2 (a in chapter 4) the slope (m in chapter 4). Note that here m is the number of points.

Fitting Polynomials in R

We will continue with the same example we begun in chapter 4, this time fitting polynomials. We can fit a polynomial instead of a straight line in **lm**. All we have to do in replace the equation **tempdiff~year** with **tempdiff~poly(year,deg)**, where **deg** is replaced by an integer giving the desired degree of the polynomial.

We begin by reading the data set and creating test and training sets.

```
temperatures=read.table("avtemp.csv", TRUE, sep=",")
years=temperatures$year
temps=temperatures$tempdiff

n = nrow(temperatures)
index = sample(1:n,round(0.75*n))
train=temperatures[index,]
test=temperatures[-index,]
```

Next, we'll do a linear fit (like we did in chapter 4 so that we have a comparison.

```
lm.fit = lm(tempdiff~year, data=train)
```

The only difference between a linear fit and a polynomial fit is using the function **poly**. Here's how we fit a degree four polynomial, for example:

```
lm.fit.4 = lm(tempdiff~poly(year,4), data=train)
```

The format is exactly the same as a linear fit except that the formula **tempdiff year** is replaced by **tempdiff poly(year,4)**.

To make predictions with polynomial we can use the function **predict**, just as we did for a linear model. We evaluate the results by calculating R values. To do this, define the *Residual Sum of Squares*

$$RSS = \sum(\hat{y}_i - y_i)^2$$

Here y_i is a test value; and $\hat{y}_i$ is a predicted value using the x_i value corresponding to that y_i; and the sum is taken over the whole test set. The RSS sums the squared differences between the predictions and observations.

Next, let

$$\overline{y} = \frac{1}{n}\sum y_i$$

and define the *Total Sum of Squares*

$$TSS = \sum(y_i - \overline{y})^2$$

The TSS sums the squared differences between the observations and the mean (average value) of all the observations.

Finally, define the *Explained Sum of Squares* as

$$ESS = \sum(\hat{y}_i - \overline{y})^2 = TSS - RSS$$

where the last step is only true for a least squares linear fit. The ESS sums the squared difference between the predictions and the average value of the observations.

Finally, we define

$$R^2 = 1 - \frac{RSS}{TSS} = \frac{ESS}{TSS}$$

Again, the second equality only holds for a linear fit. The R^2 value gives the proportion of the variation that is described by the fit, so values closer to one are best.

We can easily write a function to calculate the RSS value. The function **RSS** defined below takes three arguments: a **model**, that is the output of **lm**; a vector **observed** data (the y values in the test set); and **data**, a data frame that can be passed to **predict** to generate predictions with **model**.

```
RSS=function(model,observed,data){
    predicted=predict(model, data)
    sum((predicted-observed)^2)
}
```

To compare our linear and quartic (4th degree polynomial fit) we can calculate the RSS values of the two fits:

```
RSS(lm.fit,test$tempdiff,test)
RSS(lm.fit.4,test$tempdiff,test)
```

```
0.790469820522826
0.411262118853105
```

So the 4th degree fit seems to have a better (lower) RSS value. A normal question to ask is this: If we keep increasing the degree of the polynomial, will we keep getting a better RSS? (Note: if you try to repeat this experiment you will get a different answer because we did not see the random number generator – thus the test/train split will be different.)

We can attempt to answer this question *experimentally* by looping through the data for multiple different degrees of fit, and calculating the RSS value each time. The following code with accrue the results in an array **r2vals**

```
r2vals=0
dmax=8
r2vals[1] = RSS(lm.fit, test$tempdiff, test)
for (degree in 2:dmax) {
    model = lm(tempdiff~poly(year,degree), data=train)
    r2vals[degree]=RSS(model, test$tempdiff, test)
    }
options(repr.plot.width=7, repr.plot.height=5)

plot(1:dmax, r2vals,type="l",
    xlab="Degree",ylab="RSS")
```

Ch. 5. Polynomial Regression

We can plot the results (e.g., `plot(1:20, r2vals)` to get a plot RSS as a function of degree. The result is shown in figure 5.1.

Figure 5.1.: Plot of RSS as a function of degree of polynomial fit for range of degrees of fit from 1 (linear) to 20.

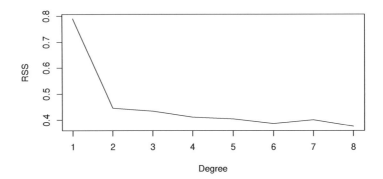

The general consensus is that there is no reason to continue adding higher terms once the RSS plots passes its "elbow." For this particular example and test/training data split, that seems to occur around $n = 2$. The remainder of the fit appears to be to noise.

```
plot(years,temps,
    xlab="Year",
    ylab="Excess Temperature",
    main="Global Average Temperature")
abline(lm.fit)
cc=sample(colors(),20)
year=1880:2017
xvals=data.frame(year)
for (degree in 2:20) {
    model = lm(tempdiff~poly(year,degree), data=train)
    ynewvals=predict(model, newdata=xvals)
    lines(year,ynewvals,col=cc[degree])
    }
```

It is useful to see what the fits actually look like. This will help us to visualize when we are under-fitting and when we are over-fitting the data. Instead of calculating the RSS values, we can just do the fits and plot them. Here we generate a random palette of 20 colors so that each line is plotted in a different color. In order to get a pretty, smooth curve for each degree, we can't send the test data to **predict**, because the points are not ordered. We could sort the data, but there is no guarantee that the data

is nicely spread in a way that will make a clean plot. Instead, we generate
an array with one point per year and that put it into a data frame with
the column name **year** so that it looks like the same variable that is in the
test data. The resulting plot is shown in figure 5.2.

Figure 5.2.: Polynomial fits of temperature data from degree 1 (linear) to 20.

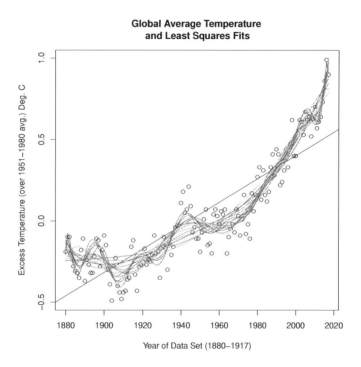

In order to understand what is going on (the math is described in the next
section) we could perform a large number test/training splits. On each
one, we should find the RSS. This removes the randomness inherent on
choosing a single split. To make the code easier, we'll write a function that
will perform find an array of RSS values. This code is specific to this large
data set (the column names are hard coded).

```
RSSVALS=function(maxdegree, train, test){
    dmax=maxdegree
    results=rep(0, maxdegree)
    model=lm(tempdiff~year, data=train)
    results[1] = RSS(model, test$tempdiff, test)
    for (degree in 2:dmax) {
        model = lm(tempdiff~poly(year,degree), data=train)
        results[degree]=RSS(model, test$tempdiff, test)
    }
    results
}
```

Then we want to wrap this function in a loop that performs multiple test/training splits. The following code will generate a sequence of 20 splits and accrue the RSS values for each one in a subsequent row of the matrix **RSSTOT**.

```
nsplits = 20
maxdeg=15
RSSTOT=rep(0,maxdeg)
for (j in 1:nsplits){
    n = nrow(temperatures)
    index = sample(1:n,round(0.75*n))
    train=temperatures[index,]
    test=temperatures[-index,]
    rss = RSSVALS(maxdeg, train, test)
    RSSTOT = rbind(RSSTOT, rss)
}
RSSTOT=RSSTOT[-1,]
```

The last line removes the row of zeros that we used to initialize the matrix. What we rally want are some statistics. We will calculate the mean and standard deviation by column. This will give us mean and standard deviation of the RSS as a function of degree.

```
means=colMeans(RSSTOT)
sds=apply(RSSTOT, 2, sd)
```

Finally, we will plot the mean +/- one standard deviation, as shown in figure 5.3. The knee still seems to be around 2. But while we see a decrease in RSS at first, eventually (for higher degree), the RSS becomes large. This is explained by the variance-bias trade-off.

```
plot(1:maxdeg, means, type="l", ylim=c(0,2))
points(1:maxdeg, means)
lines(1:maxdeg, means+sds)
lines(1:maxdeg, means-sds)
```

Figure 5.3.: RSS of fitting the temperature data from degree 1 (linear) to 20, averaged over 20 test/training splits. The upper and lower line are the one-standard deviation variation in the RSS.

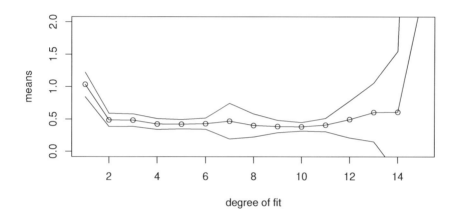

The Variance-Bias Tradeoff[3]

What is really happening can be described with probability theory. Let $y = f(x)$ be the actual function that we are sampling at some number of points, so that the true values of the function are $y_1, y_2, \ldots, y_n$ at $x_1, x_2, \ldots, x_n$. Since this function is deterministic, its expectation is $E[y_i] = y_i$. However, there is some irreducible noise in the data which we will described by its variance $\text{var}[y] = \sigma^2$. Let $\hat{y}$ be some numerical approximation (estimate) of y that we have determined from a noisy sampling of the data. What is expected difference between our estimate and the actual true function?

$$E[(y_i - \hat{y})_i^2] = E[y_i^2 + \hat{y}_i^2 - 2y_i\hat{y}_i] = E[y_i^2] + E[\hat{y}_i^2] - 2y_i E[\hat{y}_i]$$

---

[3]The mathematical content in this section can be skipped without loss of continuity.

Next, we appeal to the following result from probability theory: for any variable x, $\text{var}[x] = E[x^2] - E[x]^2$. Rearranging gives $E[x^2] = \text{var}[x] + E[x]^2$. Applying this to each of the first two terms gives

$$\begin{aligned}
E[(y_i - \hat{y})_i^2] &= \text{var}[y_i] + E[y_i]^2 + \text{var}[\hat{y}_i] + E[\hat{y}_i]^2 - 2y_i E[\hat{y}_i] \\
&= \sigma^2 + \text{var}[\hat{y}_i] + (y_i^2 - 2y_i E[\hat{y}_i] + E[\hat{y}_i]^2) \\
&= \sigma^2 + \text{var}[\hat{y}_i] + (y_i - E[\hat{y}_i])^2 \\
&= \sigma^2 + \text{var}[\hat{y}_i] + E[y_i - \hat{y}_i]^2
\end{aligned}$$

The first terms is the irreducible error, due to noise. In general, you cannot do anything to avoid this and it provides a lower limit on your fit. The second term is the variance in the model, and it describes fluctuations in your description of the data, or what you think you know about the data. As you increase the number of parameters in a model, the variance increases. The variance gives the variation in the predicted values of your model for different realizations of your model, e.g, for different training sets. The third term gives the bias. It gives the actual difference between the model and the true value you are trying to predict. The bias decreases as you add more parameters to the model, because you are improving the model. The bias varies for any given value of y_i when you change the training set and calculate a new set of parameters, so we calculate the expected value in the its definition, based on building the model a large number of times.

6. Multilinear Regression

Sometimes there will be multiple expanatory variables $x_1, \ldots, x_m$, and the dependent variable will depend on each of them linearly. To fit a model of the form
$$y = a + c_1 x_1 + c_2 x_2 + \cdots c_n x_m$$
we use the function **glm**. This will return the coefficients a and $c_1, c_2, \ldots, c_n$ for a generalized linear model.

We will use the data set **cars.csv**. This data set is based on the **auto-mpg** data set at the UCI machine learning repository, so if you want to download and reconstruct it yourself, you can.[1] We will load it as before, with **read.table**:

```
cars=read.table("~/cars.csv", TRUE, sep=",",na.strings="")
colnames(cars)
```

'mpg' 'cyl' 'displ' 'hp' 'weight' 'accel' 'year' 'origin' 'model'

The columns represent miles per gallon (**mpg**); number of cylinders in the engine (**cyl**); engine displacement in cubic inches (**displ**); vehicle weight in pounds (**weight**); acceleration (**accel**); vehicle model year (**year**); a flag that indicates the country of origin (**origin**); and a text description or name of the car (**model**). Here is a peek at the first few rows of the data frame.

mpg	cyl	displ	hp	weight	accel	year	origin	model
18	8	307	130	3504	12.0	70	1	chevrolet malibu
15	8	350	165	3693	11.5	70	1	buick skylark 320
18	8	318	150	3436	11.0	70	1	plymouth satellite
16	8	304	150	3433	12.0	70	1	amc rebel sst
17	8	302	140	3449	10.5	70	1	ford torino
15	8	429	198	4341	10.0	70	1	ford galaxie 500
14	8	454	220	4354	9.0	70	1	chevrolet impala
14	8	440	215	4312	8.5	70	1	plymouth fury iii
14	8	455	225	4425	10.0	70	1	pontiac catalina
15	8	390	190	3850	8.5	70	1	amc ambassador dpl

---

[1] The data set is at https://archive.ics.uci.edu/ml/machine-learning-databases/auto-mpg/. It is not stored as CSV file and there is missing data which is marked with question marks.

We'll begin as before, separating the data into test and training sets. Well use 75% for training. Since some of the data is missing, we'll edit out that data using **na.omit**.

```
okcars=na.omit(cars)
n=nrow(okcars)
ind = sample(1:nrow(cars),round(0.75*n))
train=data[index,]
test=data[-index,]
```

Since we are only trying to fit the quantitative variables, our model should have the form

$$(\text{mpg}) = a + b \times (\text{cyl}) + c \times (\text{displ}) + d \times (\text{weight}) + e \times (\text{accel})$$

where a, b, c, d, and e are undetermined coefficients. The call to **glm** is similar to the call to **lm**.

```
mpg.fit = glm(mpg~cyl+displ+hp+weight+accel, data=train)
mpg.fit
```

```
Call:  glm(formula = mpg ~ cyl + displ + hp + weight + accel,
       data = train)

Coefficients:
(Intercept)          cyl         displ            hp       weight
accel
   48.711775    -0.398889      0.003724     -0.052935    -0.005724
   -0.069353

Degrees of Freedom: 293 Total (i.e. Null);  288 Residual
Null Deviance:       18460
Residual Deviance: 5157         AIC: 1691
```

It was not necessary to print the results of the fit, but it is useful to see what the results are, and it gives us an idea of how much the miles per gallon depends on each variable. The coefficients row tells us that the fit requession equation is

$$\text{mpg} = 48.711775 - 0.398889(\text{cyl}) + 0.003724(\text{displ}) - 0.052935(\text{hp}) \\ - 0.005724(\text{weight}) - 0.069353(\text{accel})$$

The largest slope, for example, is the number of cylinders: for each additional cylinder the car has, the model predicts decrease of by nearly 0.4

mpg. However, the strength of the dependence is not the same as the goodness of fit of the model. This can be explained by the RSS or RMS error. To avoid typing in the formula, lets define a function **rms(obs,pred)** that takes two vectors of the same length and returns the rms error:

```
rms = function(obs,exp){
    n=length(exp);
    sqrt((sum((obs-exp)^2))/n)}
```

The indentation is not required; the entire function could be written on a single line if desired. This format is better because it is easier to read. Note that this function does not do any error checking, so it will crash and burn if the vectors have different lengths.

The RMS error of the fit is then determined by performing a prediction.

```
yobs=test$mpg # use $mpg to extract column mpg
multihat=predict(mpg.fit,newdata=test)
rms(multihat,yobs)
```

4.35449277408379

We can perform a one dimensional fit of

$$(\text{mpg}) = a + b \times (\text{weight})$$

using

```
mpg.weight.fit = glm(mpg~weight, data=train)
weighthat=predict(mpg.weight.fit,newdata=test)
rms(weighthat,yobs)
```

4.44190910861038

Similarly, we can find a one dimensional fit of

$$(\text{mpg}) = a + b \times (\text{cyl})$$

```
mpg.cyl.fit = glm(mpg~cyl, data=train)
cylhat=predict(mpg.cyl.fit,newdata=test)
rms(cylhat,yobs)
```

4.62535034333826

Similarly, we can find a one dimensional fit of
$$(\text{mpg}) = a + b \times (\text{hp})$$

```
mpg.hp.fit = glm(mpg~hp, data=train)
hphat=predict(mpg.hp.fit,newdata=test)
rms(hphat,yobs)
```

4.79093400664866

Finally, we can perform a one dimensional fit of
$$(\text{mpg}) = a + b \times (\text{accel})$$

```
mpg.accel.fit=glm(mpg~accel,data=train)
accelhat=predict(mpg.accel.fit, newdata=test)
rms(accelhat,yobs)
```

6.74651299042712

We could even repeat a simple polynomial regression against the single variable weight,

```
mpg.poly.weight.fit = lm(mpg~poly(weight,3), data=train)
poly.weighthat=predict(mpg.poly.weight.fit,newdata=test)
rms(poly.weighthat,yobs)
```

4.55984434159997

As we see, the RMS error is still better for the full multilinear model than for either the cubic fit or any of the individual linear models.

Since the data is multidimensional, it is difficult to visualize. Usually we look at projections onto the most important planes, for example, **mpg** vs **weight**. In the following plot the full fit is shown as a thick solid line, while a simple linear regression against weight is shown as a thinner dashed line. The polynomial regression is shown as the curve.

```
plot(cars$weight,cars$mpg,cex=.25,
    xlab="Weight of Car in Pounds",
    ylab="Miles Per Gallon (mpg)",main="Least Squares Fit")
abline(simple.linear.regression, lty="dashed")
lines(test$weight,
    weighthat,col="red",lwd=4) #$pulls out weight column
lines(weight, for.plot.poly.y,
    col="blue", lwd=3, lty="dashed")
```

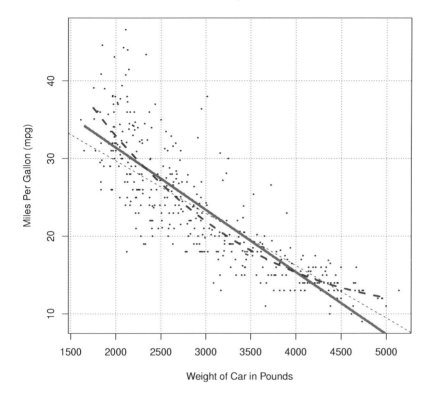

The function **pairs** can be used to plot a grid of all coordinate projections.

```
pairs(mpg~weight+cyl+hp+accel, data=cars, cex=.1)
```

The output is shown in figure 6.1.

Using lm or glm?

You may have noticed that we slipped in the function **glm** in place of the linear model **lm**. Both functions do basically the same thing, with one notable exception:

- **lm** assumes that the noise is distributed normally (also called a Gaussian noise model)

- **glm** allows you to specify a noise distribution

We did not take advantage of that in the discussion above. To select a

Figure 6.1.: Output of **pairs** command for the **mpg** data.

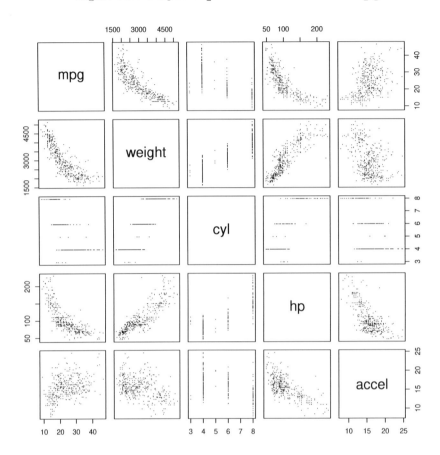

model, you would have to specify `family=gaussian`, where `gaussian` is replaced by one of values in table 6.1.

Table 6.1. Noise Models For Linear Models[1]		
`binomial`	`gaussian`	`Gamma`
`inverse.gaussian`	`quasi`	`quasibinomial`
	`quasipoisson`	

[1]These are possible values for the `family` option to `glm`. For details see https://www.rdocumentation.org/packages/stats/versions/3.5.0/topics/family.

The Math of Generalized Linear Regression[2]

Suppose we have p observations of an n-dimensional data set that we write as row vectors

$$y_1 = [x_1, x_2, \ldots, x_n]_1$$
$$y_2 = [x_1, x_2, \ldots, x_n]_2$$
$$\vdots$$
$$y_p = [x_1, x_2, \ldots, x_n]_p$$

We want to approximate these observations by the best fit plane

$$y = c_0 + c_1 x_1 + c_2 x_2 + \cdots + c_n x_n$$

using the method of least squares. The observations tell us that a perfect fit would give

$$c_0 + c_1 x_{11} + c_2 x_{21} + \cdots + c_n x_{n1} = y_1$$
$$c_0 + c_1 x_{12} + c_2 x_{22} + \cdots + c_n x_{n2} = y_2$$
$$\vdots$$
$$c_0 + c_1 x_{1p} + c_2 x2p + \cdots c_n x_{np} = y_p$$

where the first index refers to the vector component and the second index to the observation number. This can be rewritten as the matrix equation

$$\begin{bmatrix} 1 & x_{11} & x_{21} & \cdots & x_{n1} \\ 1 & x_{12} & x_{22} & \cdots & x_{n2} \\ \vdots & & & & \vdots \\ 1 & x_{1p} & x_{2p} & \cdots & x_{np} \end{bmatrix} \begin{bmatrix} c_0 \\ c_1 \\ \vdots \\ c_n \end{bmatrix} = \begin{bmatrix} y_0 \\ y_1 \\ \vdots \\ y_n \end{bmatrix}$$

Defining the matrix on the left as **A** and the two column vectors as **c** and **y**, any real solution will have some residual error **r**, where

$$\mathbf{Ac} = \mathbf{y} + \mathbf{r}$$

The goal is to find the values of the vector **c** that minimize the residual distance

$$|\mathbf{r}| = |\mathbf{AC} - \mathbf{y}|$$

---

[2] This section can be skipped without loss of continuity. The material in this section is based on the author's text *Scientific Computation*, 3rd ed.

We have already solved this problem (with a different matrix **A**) for polynomial least squares (see equations 5 through ??). The solution is

$$\mathbf{A}^\mathrm{T}\mathbf{y} = \mathbf{A}^\mathrm{T}\mathbf{A}\mathbf{c}$$

The matrix $= \mathbf{A}^\mathrm{T}\mathbf{A}$ is symmetric and hence invertible, and therefore

$$\mathbf{c} = (\mathbf{A}^\mathrm{T}\mathbf{A})^{-1}\mathbf{A}\mathbf{y}$$

If we define the column vectors $\mathbf{u}_j^\mathrm{T} = [1, x_{1j}, x_{2j}, x_{3j}, \ldots, x_{nj}], j = 1, 2, \ldots, p$, then

$$\mathbf{A}^\mathrm{T}\mathbf{A} = \begin{bmatrix} \mathbf{u}_1 | \cdots | \mathbf{u}_p \end{bmatrix} \begin{bmatrix} \mathbf{u}_1^\mathrm{T} \\ \mathbf{u}_2^\mathrm{T} \\ \vdots \\ \mathbf{u}_p^\mathrm{T} \end{bmatrix} = \begin{bmatrix} p & \sum x_{1i} & \sum x_{2i} & \cdots & \sum x_{ni} \\ \sum x_{1i} & \sum x_{1i}^2 & \sum x_{1i}x_{2i} & \cdots & \sum x_{1i}x_{ni} \\ \sum x_{2i} & \sum x_{2i}x_{1i} & \sum x_{2i}^2 & \cdots & \vdots \\ \vdots & & & & \vdots \\ \sum x_{ni} & \cdots & & & \sum x_{ni}^2 \end{bmatrix}$$

If the observations have three variables $[x_i, y_i, z_i]$, then

$$\mathbf{A}^\mathrm{T}\mathbf{A} = \begin{bmatrix} p & \sum x_i & \sum y_i & \sum z_i \\ \sum x_i & \sum x_i^2 & \sum x_i y_i & \sum x_i z_i \\ \sum y_i & \sum y_i x_i & \sum y_i^2 & \sum y_i z_i \\ \sum z_i & \sum z_i x_i & \sum z_i y_i & \sum z_i^2 \end{bmatrix}$$

Ridge and Lasso Regression

One of the common problems with models of the form

$$\hat{y} = a + \mathbf{c}^\mathrm{T}\mathbf{x}$$

is that of **multicollinearity**. When multicollinearity occurs (or nearly occurs) it is possible to predict one (or more) explanatory variable from the others. Thus the different dimensions in **x** may be highly correlated. The solution proposed by least squares is to find a predictor that minimizes

$$\mathcal{E} = \sum_{i=1}^n |y_i - a_i - \mathbf{c}_i^\mathrm{T}\mathbf{x}_i|^2$$

where the sum is over all the training points. In **ridge regression**, a method originally formulated to for the solution of integral equations[3]

solves this problem by constraining the least square solution to a space

$$\sum_{i=1}^{n} c_i^2 \leqslant K$$

where the constant K is a parameter of the problem. Equivalently, the problem can be solved by minimizing

$$\tilde{\mathcal{E}} = \sum_{i=1}^{n} \left| y_i - a_i - \mathbf{c}_i^{\mathrm{T}} \mathbf{x}_i \right|^2 + \lambda \sum_{i=1}^{n} c_i^2$$

Ridge regression can be performed using the function `lm.ridge` in the **MASS** package.[4]

An alternative to ridge regression is **lasso regression**.[5] The lasso replaces the sum-of-squares of the coefficients used in ridge regression with the sum of absolute values, but is otherwise conceptually the same. The energy function is minimized subject to the constraint

$$\sum_{i=1}^{n} |c_i| \leqslant K$$

or equivalently, to minimize the modified energy function

$$\tilde{\mathcal{E}} = \sum_{i=1}^{n} \left| y_i - a_i - \mathbf{c}_i^{\mathrm{T}} \mathbf{x}_i \right|^2 + \lambda \sum_{i=1}^{n} |c_i|$$

The lasso is implemented in the R package **glmnet**.[6] For instructions, see the vignette at CRAN.[7]

---

[3] Tikhonov, A. N. (1943). [On the stability of inverse problems (in Russion)]. Doklady Akademii Nauk SSSR. 39:195-198

[4] Ripley, B, et. al. (1998) *MASS: Support Functions and Datasets for Venables and Ripley's MASS* . Software and documentation are available at **MASS** package is available at https://cran.r-project.org/package=MASS

[5] Tibshirani, R. (1996). *Regression Shrinkage and Selection via the lasso*. Journal of the Royal Statistical Society. Series B (methodological). 58: 267-88.

[6] Friedman, et. al. (2010). *Regularization Paths for Generalized Linear Models via Coordinate Descent.* Journal of Statistical Software, 33:1-22. Software available at CRAN: Friedman, J. et.al. (2018) *glmnet: Lasso and Elastic-Net Regularized Generalized Linear Models,* https://cran.r-project.org/package=glmnet

[7] Available at https://cran.r-project.org/web/packages/glmnet/vignettes/glmnet_beta.pdf

7. Nonlinear Regression

The goal of least squares nonlinear regression is the same as linear regression: to find the best fit to a function. We want to find the parameters for some $\hat{y} = \hat{f}(x; a_1, \dots)$ such that the error function

$$\mathcal{E} = \sum_{i=1}^{n} |\hat{y}_i - y_i|^2 = \sum_{i=1}^{n} |\hat{f}(x_i; a_1, a_2, \dots) - y_i|^2$$

is minimized. The problem in nonlinear regression is that when the partial derivatives with respect to the unknown parameters $a_1, a_2, \dots$ are calculated, they are not linear in $a_1, a_2, \dots$. Hence it is rarely possible to find explicit analytic formulas for the solutions. An example of a nonlinear function that is commonly fit to data in chemistry and biology is the **Hill** function. A Hill function has the form

$$y = \frac{x^n}{a + x^n}$$

Here the unknown parameters are a and n, where $a > 0$ and n can either positive or negative. Typically a Hill function occurs in an n-step reaction and gives the relative rate at which a reaction occurs, but there are also situations where non-integral values of n will occur.

Instead of the matrix formulation used in linear or polynomial regression, a numerical iteration is required. The most commonly used procedures are generally variations in **gradient descent**. In these methods, a first guess (often supplied by the user) is made for the parameters. This could be as simple as setting all unknowns equal to 1. Then partial derivatives $\partial \mathcal{E}/\partial a_i$ are calculated with respect to all unknown parameters. An incremental update to all parameters is made that takes the value of $\mathcal{E}$ **downhill**, i.e., in such a way so as to decrease its value. If no such value can be found that $\mathcal{E}$ is already at a local minimum. The process is then repeated iteratively until $\mathcal{E}$ cannot be decreased any further or until a convergence criterion is met.

Unfortunately gradient descent converges very slowly; numerous workarounds have been implemented to improve the rate of convergence and these are the subject of considerable interest in computational science.See, for example, Strang, G (2007) *Computational Science and Engineering*, Wellesley-Cambridge Press. Since the energy function may not converge to the global minimum there is always the possibility of finding a non-optimal solution, so it is best to use minimization functions that are already implemented and have considered this issue. Fortunately, many computer languages (R included) have nonlinear solvers already built into them, so it is not necessary to actually perform the minimization. The solver figures out the necessary minimization technique for you and gives the solution.

We will demonstrate the technique by fitting a hill function in R using the nonlinear solve **nls**. Before we do this, we will generate some randomized test data. Since we want the targeted data to look like a hill function, we will define a simple hill function.

```
f=function(x,a=1,n=1){(x^n)/(a+x^n)}
```

Next, we will produce the **target** curve. This is not the data we are fitting. The target curves is a Hill function (used primarily for pretty-plotting) that will help us to place the random numbers. We'll generate a target function with $n = 2.5$ and $a = 1.25$. Then we'll produce arrays of closely spaced points in **xplot** and **yplot** for later pretty-plotting using these parameters.

```
xplot=seq(0,5,by=0.01)
n.target=2.5
a.target=1.25
yplot=f(xplot,a.target,n.target)
```

Now we want to generate the pseudo-experiment. These will be data points that are scattered around the curve. There will not be as many of them. For the plot we had points separated by $\Delta x = 0.01$; here we will separate the points along the x axis by 0.2. We'll scatter our points about the target curve by adding in a noise element that has mean of zero and standard deviation of 0.1.

```
x=seq(0,5,by=0.2)
n=length(x)
y=f(x, a.target, n.target)+rnorm(n,0,.1)
```

Our training data is now in arrays **x** and **y**. We are ready to do a nonlinear fit. We'll use **a=1** and **n=1** as starting values.

```
model=nls(y ~ x^n/(a+x^n),start=list(a=1,n=1))
print(model)
```

```
Formula: y ~ x^n/(a + x^n)

Parameters:
  Estimate Std. Error t value Pr(>|t|)
a   0.8887     0.1355   6.560 8.75e-07 ***
n   2.3743     0.2996   7.925 3.72e-08 ***
---
Signif. codes:  0 '***' 0.001 '**' 0.01 '*' 0.05 '.' 0.1 ' '

Residual standard error: 0.08745 on 24 degrees of freedom

Number of iterations to convergence: 8
Achieved convergence tolerance: 5.067e-06
```

Figure 7.1.: Fit to a Hill function using **nls**. Markers give the randomized data points used in the fit. The thicker, lighter line is the predicted fit; the thin, solid line is the target curve.

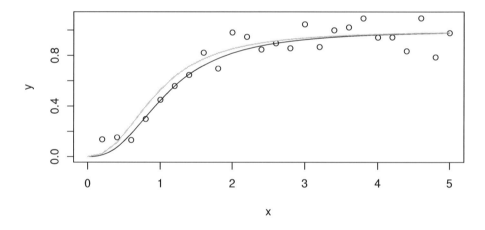

Extract the parameters; its just as easy to do this as to run a **predict** function.

```
parameters=summary(model)$param    # $parameter attribute
print(parameters)
a.fit=parameters[1]
n.fit=parameters[2]
print(c(a.fit,n.fit))
```

```
         Estimate Std. Error   t value     Pr(>|t|)
a 0.8887127  0.1354830 6.559590 8.752050e-07
n 2.3743297  0.2995851 7.925394 3.723794e-08
[1] 0.8887127 2.3743297
```

Generate and plot predictions.

```
prediction=f(x,a.fit,n.fit)
options(repr.plot.width=7, repr.plot.height=5)
plot(xplot, yplot,type="l", ylab="y",xlab="x", ylim=c(0,1.1))
points(x,y)
lines(x, prediction, lwd="1.5",col="gray")
```

The resulting prediction, target curve, and randomized data are plotted in figure 7.1.

Ch. 7. Nonlinear Regression

8. Backprop Networks

Neural networks can be used for both regression and classification. Many different types of neural networks have been invented, ranging from the earliest perceptron to the latest deep learning networks. We will discuss using one type of neural network for regression in this chapter and return to using neural networks for classification again in later chapters.

Background

Artificial neural networks (ANN's) are built by combining single **perceptrons**. The perceptron [1] is sometimes called the McCulloch Pitts neuron in honor of its inventors.[2] At its simplest, the perceptron linearly combines inputs $x_1, x_2, \ldots, x_n$ with weighting factors $w_1, w_2, \ldots, w_n$ (figure 8.1). The inputs and weights can be combined into vectors **v** and **w** to give the equation

Figure 8.1.: A single neuron perceptron without a thresholding function.

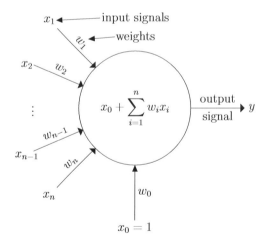

---

[1] Rosenblatt, F (1957) *The Perceptron–a perceiving and recognizing automaton.* Report 85-460-1, Cornell Aeronautical Laboratory

[2] McCulloch, W., Pitts, W. (1943) *A logical calculus of the ideas immanent in nervous activity.* Bulletin of Mathematical Biophysics, **5**:115-133.

$$y = w_0 + \mathbf{w} \cdot \mathbf{x} = \sum_{k=0}^{n} w_i x_i$$

Here we have arbitrarily extended the vector by one by adding an extra input that always has strength $x_0 = 1$. In two dimensions (with one input) this is the equation of a straight line with slope w_1 and intercept w_0. In more dimensions, this is the equation of a hyperplane.

With a suitable choice of weights the perceptron can be used to find a hyperplane that linearly separates the data in two sets. These plane can either represent a best fit (linear regression) or separation (for classification).

However, the neuron as defined here is deterministic. Real neurons have a certain probability of firing (producing an output) that is determined by the input. To represent this probability of a neuron firing, we apply a probability distribution σ to y:

$$y = \sigma(w_0 + \mathbf{w} \cdot \mathbf{x})$$

Since σ is a probability distribution is must be increasing from zero to 1, and saturate at $y = 1$ as x becomes large (figure 8.2). In order to make this work, all of the inputs are normalized to the interval $[0, 1]$. Typical thresholding functions are signmodial (such as the logistic sigmoid $g(x) = 1/(1 + e^{-x})$, and the unit step function). The idea of a thresholding function is to turn on the output when the input is large and to turn it off when the input is small.

Individual neurons can be combined together into a network called a single layer perceptron (figure 8.3). This will lead to a linear separation in the data because of the nature of the thresholding function and the format of the argument $w_0 + \mathbf{w} \cdot \mathbf{x}$. When $w_0 + \mathbf{w} \cdot \mathbf{x} > 0$ then the threshold function will cause the neuron to fire its output; when $w_0 + \mathbf{w} \cdot \mathbf{x} \leq 0$, the neuron will not fire its output. At the boundary, we have $w_0 + \mathbf{w} \cdot \mathbf{x} = 0$ which is the equation of a straight line.

Lets look at how the single layer perceptron works. Suppose that there are n output neurons $\mathbf{y} = (y_1, \ldots, y_n)$ and p input neurons $\mathbf{x} = (1, x_1, \ldots, x_p)$. (In figure 8.1 we $n = 3$ and $p = 4$.) We have no idea what weights to use, so we use some very small random numbers to initialize all the weights

Figure 8.2.: A single neuron perceptron with a thresholding function.

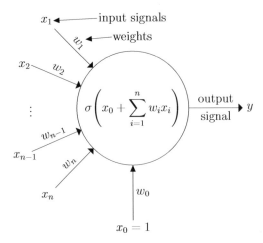

w_{ij}. Using the given input and initial weight values, we will calculate an observed output $(y_1, \ldots, y_n) = \mathbf{y} = \sigma(\mathbf{w} \cdot \mathbf{x})$. During training we will have some training values $\mathbf{y'} = (y'_1, \ldots, y'_n)$, which will be different from the actual values $\mathbf{y} = (y_1, \ldots, y_n)$. we will use this error to update the the vector of weights by

$$\Delta \mathbf{w} = \eta(\mathbf{y'} - \mathbf{y}) \cdot \mathbf{x}$$

where η is a very small number. The process is then repeated with the new weights and the iteration is repeated until the algorithm converges.

This nonlinear transformation is crucial to the combination of single neurons to form larger neural networks. The idea is to combine multiple neurons by taking the outputs of some neurons and feeding them into other neurons. In a **multilayer perceptron** the neurons are arranged in layers as a feed-forward network (figure 8.4). If we only had linear combinations, adding more neurons would not do anything for us, since the linear combination of a linear combination is still linear. But by adding nonlinearity it turns out that we can fit (and separate) practically any type of reasonably shaped function. Each neuron processes its input by calculating the weighted sum of the extended vector (with weight w_0) and then applies the threshold function σ.

The update rule for a multilayer perceptron uses the back-propagation rule.

Figure 8.3.: A single layer perceptron with $p = 4$ inputs and $n = 3$ outputs. There is only a single layer. The node labels $\sigma(\sum)$ are shorthand for $y_i = \sigma(\sum_{k=1}^{p} w_{ik} x_k)$. Each arrow has a different weight w_{ik} associated with it, where the first index refers to the output node and the second index refers to the input node.

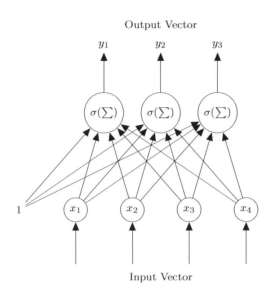

Derivation of Back-Propagation Rule[3]

We update the errors by finding weights that minimize the sum squared errors,

$$\mathcal{E} = \frac{1}{2} \sum (y'_i - y_i)^2$$

where the sum is taken over the training data. Here y'_i is a value in the training data and y_i is a value in the output vector. We note that this is the same sum we minimized when solving for the coefficients in linear regression. The leading factor of $1/2$ is not necessary, but is added for convenience, to cancel out the factor of 2 that arises during differentiation. u_i be the output of any neuron in any given layer. The if that neuron has p inputs $u_1, \ldots, u_p$ from the underlying layer,

$$u_j = \sigma \left(\sum_{k=1}^{p} w_{ki} u_k \right)$$

---

[3]This section can be skipped without loss of continuity

Figure 8.4.: A multilayer perceptron may have zero or more hidden layers. This figure illustrates a multilayer perceptron with four inputs, 3 outputs and a five cell hidden layer.

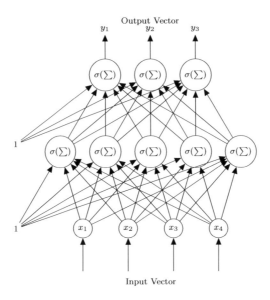

To find the optimal w_{ij} we need to solve for the partial derivatives

$$\frac{\partial \mathcal{E}}{\partial w_{ij}} = \frac{\partial \mathcal{E}}{\partial u_j} \frac{\partial u_j}{\partial w_{ij}}$$

If u_j is an output layer, then $u_j = y_j$, so

$$\frac{\partial \mathcal{E}}{\partial u_j} = \frac{\partial}{\partial y_j} \frac{1}{2} \sum (y'_i - y_i)^2 = y'_j - y_j$$

If we take $\sigma(x)$ to be the logistic function $\sigma(x) = 1/(1 + e^{-x})$ then

$$\frac{\partial \sigma(x)}{\partial x} = \frac{\partial}{\partial x} \frac{1}{1 + e^{-x}} = \frac{e^{-x}}{(1 + e^{-x})^2} = \frac{e^{-x}\sigma(x)}{1 + e^{-x}}$$

But

$$1 - \sigma(x) = 1 - \frac{1}{1 + e^{-x}} = \frac{e^{-x}}{1 - e^{-x}}$$

and thus

$$\frac{\partial \sigma(x)}{\partial x} = \sigma(x)(1 - \sigma(x))$$

Using the fact that $u_j = \sigma\left(\sum_{k=1}^{p} w_{ki} u_k\right)$, for the output layer

$$\frac{\partial u_j}{\partial w_{ij}} = \frac{\partial}{\partial w_{ij}} \sigma\left(\sum_{k=1}^{p} w_{ki} u_k\right) = u_j(1-u_j) \frac{\partial}{\partial w_{ij}} \sum_{k=1}^{p} w_{ki} u_k$$
$$= u_j(1-u_j) u_i$$

Thus (for the output layer)

$$\frac{\partial \mathcal{E}}{\partial w_{ij}} = y_j(1-y_j)(y'_j - y_j) u_i$$

There is a more complicated formula for the hidden layers. The reader is referred to more advanced books for a derivation. The resulting learning rule is

$$\Delta w_{ij} = -\eta \delta_j y_i$$

where

$$\delta_j = \begin{cases} y_j(1-y_j)(y'_j - y_j) & \text{for the output layer} \\ y_j(1-y_j)(\sum \delta_k w_{jk}) & \text{for hidden layers} \end{cases}$$

ANN's for Linear Regression

Here we will build a simple one-layer linear network for the miles per gallon data file from chapter 6. Since we are fitting a multi-linear model we don't expect to do any better than we did there.

We start by loading the necessary library.[4]

```
install.packages("neuralnet")
library(neuralnet)
```

Next, read the data and print the column names. Since this file may contain missing data, we want to edit this data out with the **na.omit** function.

```
options(width=70) # set line wrap to 70 chars
cars=read.table("~/cars.csv", TRUE,
     sep=",",na.strings="")
okcars=na.omit(cars) # rows lines with missing data
print(colnames(okcars))
```

---

[4] Documentation of the **neuralnet** package is at https://cran.r-project.org/web/packages/neuralnet/.

```
[1] "mpg"    "cyl"   "displ"  "hp"   "weight"  "accel"  "year"
[8] "origin" "model"
```

After we read the file, we need scale all the variables to the unit interval. To do this we subtract the column minimum from each variable and then divide the result from the column range:

$$\tilde{y}_{i,j} = \frac{y_{i,j} - y_{j,\min}}{y_{j,\max} - y_{j,\min}}$$

where $y_{j,\max}$ is the maximum value of column j, and $y_{j,\min}$ is the minimum value in column j. The **scale** function will do this. The general format is

```
scale(dataframe,
      center=column.centers,
      scale=column.widths)}
```

where **column.center** and **column.widths** are vectors that have the same length as the number of columns in **dataframe**. The first vector contains the desired centers for scaling, and the second vectors contains the desired widths for scaling. They must be in the same order as the columns in the data frame. If the data frame contains non-numeric data this will cause an error.

We can use the **apply** function to evaluate a function on one or more rows or columns of a data frame. The format is

```
apply(dataframe, code, f)
```

If **code** equals 1, then **f** is applied to each row of the data frame. If **code** equals 2, then **f** is applied to each column of the data frame.

We want to use **apply** to calculate the minimum and maximum value of columns 1 through 6 of the data frame **okcars** using the functions **max** and **min** respectively. We will use the function **scale** just discussed to generate a matrix of scaled values, and then we will convert it back to a data frame using **as.data.frame**.

```
maximums = apply(okcars[1:6], 2, max)
minimums = apply(okcars[1:6], 2, min)
widths=maximums - minimums
scaled_cars = as.data.frame(scale(okcars[1:6],
   center = minimums, scale = widths))
```

Now we can produce training and test set. Reserving 75% for test as before,

```
n=nrow(okcars)
ind=sample(1:n,round(.75*n))
train=scaled_cars[ind,]
test=scaled_cars[-ind,]
```

We generate the model with the function **neuralnet**:

```
nn.model = neuralnet(mpg ~ cyl+displ+hp+weight+accel,
   data=train,
   hidden=c(3),
   linear.output=T)
```

The option **hidden** is a vector that says how many nodes we want in each hidden layer. Here we are requesting a single hidden layer with three nodes. The option **linear.option** tells **neuralnet** that we are trying to learn a linear model.

Once we have the model, we make predictions using the test set. The option **test** should be a matrix-like object with columns in the same order as the model. The second line scales the data back to MPG from scaled values.

```
nn.predictions = compute(nn.model,test[2:6])
pre=nn.predictions$net.result*widths[1]
   +minimums[1]  # $net.results has values
obs=test_neuralnet[,1]*widths[1]+minimums[1]
R2=1-sum((obs-pre)^2)/sum((obs-mean(obs))^2)
```

```
0.625990835885157
```

As we see from the R^2 value, the result is comparable to the regression model. by some tuning of model features (number of nodes) we should be able to get an exact match.

The output of **neuralnet** is compatible with **plot**, and a visualization of the network will be displayed.

```
plot(nn.model,
   show.weights=TRUE, # flag for displaying weights
   rep="best",        # which iteration to display
   radius=.1,         # radius of nodes
   arrow.length=.25)  # length of arrows
```

The output is shown in figure 8.5. As annotated in the figure, the model required 580 iterations to converge. If you do not specify `rep=``best''` in the `plot` statement, the function will attempt to plot all 580 iterations (and probably fail by running out of video memory).

Figure 8.5.: Neural network found for the miles per gallon data. A linear model was fit and there is a single hidden layer with three cells.

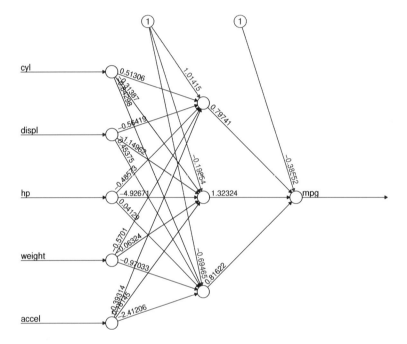

It is actually rather difficult to visualize the prediction meaningfully, even though the model is linear in all of the variables. For example, we could try to plot the predicted MPG as a function of vehicle weight with all other variables held fixed. But the problem with this is in choosing what to fix the other values at. if we fix the values at their averages, for example, we would end up treating all cars as having around 5.5 cylinders. But since the model is also variable with other factors, it is not meaningful to treat 2000 pound cars and 5000 pound cars as having the same numbers of cylinders and same engine displacements.

One way to look at this is to group the data into bins, of say 500 pounds.

Say we pick all the cars between 2000 and 2500 pounds, and find the average of all their parameters. Use the averages as the input to the predictor, and see what the model suggests. This gives one point, which we can plot at the (x-value) weight average and predicted mpg (y-value). Then repeat for the 2500-3000 weight class, and so forth. If most cars in each weight class are fairly similar, this will give a reasonable fix of the non-weight parameters.

Here's how we might do the binning. First, create an empty data frame with column headers that have the same names and types as the model.

```
df=data.frame(cyl=double(), displ=double(), hp=double(),
    weight=double(),accel=double())
```

Then for each bin of width 500 pounds, with left end points at 500, 1000, 1500, ... 4500 pounds, extract those rows of the data set that are in the desired weight class using **subset**. Use **apply** to calculate the means by column. Then append the row to the end of the data frame using **bind**. Since building a data frame this way destroys the names, we have to rename all the columns at the end.

```
for (bin in seq(1500, 5000, by=500)){

    # extract all rows in the weight class
    therows=subset(okcars[,1:6],weight>=bin & weight<bin+500)

    # skip to next iteration if nothing found
    if (length(therows)<1) {next}

    # find mean of each column
    rowmeans=apply(therows,2,mean)[2:6]

    # append the new list of means to df
    df=rbind(df,rowmeans)

}
# give names to the columns
colnames(df)=c("cyl","displ","hp","weight","accel")
format(df,digits=2)
```

cyl	displ	hp	weight	accel
4.0	88	64	1892	17
4.0	104	78	2222	16
4.6	150	95	2738	16

Ch. 8. Backprop Networks

6.2	234	105	3262	16
7.2	293	129	3735	15
8.0	349	163	4262	13
8.0	382	178	4752	13
8.0	400	175	5140	12

The last line let us peek into the results. The **format** function here is only used for printing the result. It allows us to round off everything in the file to two digits before printing so that it is easier to read.

Now that we have a data frame with the test data averaged into bins, we can scale it using the same minimum and width values as we did before.

```
scaleddf = as.data.frame(scale(df, center = minimums[2:6],
    scale = widths[2:6]))
```

Here we use **compute** to make a prediction. We use the scaled data frame we just computed as our input data. To convert the result back to MPG we must extract the **net.result** column of the output of **compute** and rescale it. This will become the y value of the predicted plot. For the x value we use the **weight** column of the unscaled data frame.

```
nn.predict.for.plot = compute(nn.model,scaleddf)
yvals=nn.predict.for.plot$net.result*widths[1]+minimums[1]
xvals=df$weight

# make a scatter plot of raw data
plot(cars$weight,cars$mpg, pch=5,cex=.5,
     grid(lty="dotted",col="black"),
     xlab="Vehicle Weight", ylab="Miles Per Gallon")

# make a line plot of predicted data
lines(xvals,yvals)
```

The line plot (figure 8.6) of the predicted data has one point per 500 pound bin. The point is plotted at an x value based on the average weight of all the cars in the bin weight class (which is generally not the center of the bin). The y value is the predicted value from **compute**, rescaled to MPG units.

Figure 8.6.: The markers show the raw data of the car MPG data set. The line connects the predictions of the neural net, where the input data used for the plot are the bin averages of 500 pound car weight classes. If all variables except for mpg and weight were held fixed, the prediction would be a straight line. It is not a straight line because a different set of fixed variables are used as input for each 500 pound weight class. This ultimately reflects the overall concave up, decreasing shape of the data.

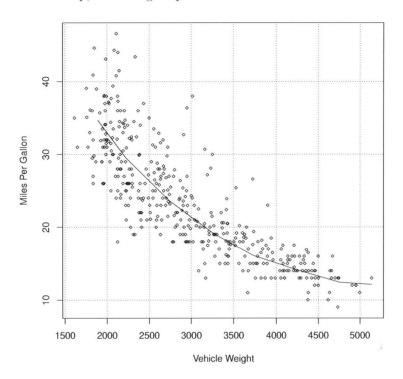

Using Neural Networks to Fit A Nonlinear Time Series

The 10.7 centimeter solar flux, normally abbreviated as $F_{10.7}$ or F10.7, is a measure of the average solar energy incident upon the earth (figure 8.7). It is called the 10.7cm flux because it is actually a measure of solar radiation (light) at the 10.7cm wavelength (which is in the microwave range of the spectrum). Variations in the solar flux are correlated with solar activity (such as sunspots) and have 28 day, 11 year, and 22 year cycles.[5] This

---

[5] The 28 day cycle corresponds to one solar rotation. The cause of the 11 and 22 year cycles is unknown.

variability is important because it (a) affects upper atmospheric density[6] and (b) impacts the ability of satellites to operate properly. Solar flux is measured in solar flux units (SFU), where one SFU is equal to 10^{-22} Joules/m^2.

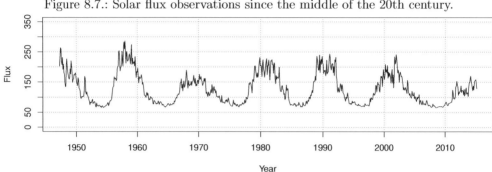

Figure 8.7.: Solar flux observations since the middle of the 20th century.

Unfortunately, there are no accurate physical models of solar activity so all predictive models are empirical. The problem is compounded by the fact that sunspots (like an earthquake or volcanic eruptions) often appear unexpectedly, seemingly at random. We can attempt to fit the periodic data but there is little we can do to predict sudden random changes. It is likely that solar activity is also connected to terrestrial weather. The sun is currently in relatively high state of activity – higher than it has been for about 10,000 years. The current period of high activity is not sufficient to explain the overall global warming that has occurred over the past century.

There have been brief periods of reduced activity during recorded history with corresponding decreases in temperature in the northern hemisphere (9th C BC, 1040-1180, 1650-1710, and 1790-1820. The last minimum corresponded to a period of significant volcanic eruption in (Mt. Tambora, in Indonesia) in 1815 that was followed by the "year without a summer" and widespread famine in North America and Europe. Whether or not the impact of the Tambora eruption would have been as significant at higher flux levels is unknown.

We will fit a nonlinear neural network to this data.

First, read the data set.

---

[6]At higher flux levels (which are correlated with higher sunspot activity), the atmosphere expands due to thermal warming, increasing drag.

```
solarflux=read.table("~/solar-flux.csv", TRUE,
    sep=",",na.strings="")
colnames(solarflux)
```

`'date' 'flux'`

The data set only has two columns, **date** and **flux**. Our network will only have one input node, and several input layers.

Separate the data into test and training sets. Before we make our selections, we will sort the indices to keep the sets in time order to make later plotting easier.

```
n=nrow(solarflux)
ind=sort(sample(1:n,round(.75*n)))
```

Scale the data as we did for the linear fit.

```
maximums = apply(solarflux[1:2], 2, max)
minimums = apply(solarflux[1:2], 2, min)
widths=maximums - minimums
scaled_flux = as.data.frame(scale(solarflux[1:2],
    center = minimums, scale = widths))
```

Now we are ready to partition the **scaled_flux** into test and training sets.

```
train=scaled_flux[ind,]
test=scaled_flux[-ind,]
```

Peek at the training set:

```
train[1:10,]
```

	date	flux
5	0.004897058824	0.6881924660
6	0.006132352941	0.6415683082
7	0.007352941176	0.7500113056
8	0.008573529412	0.5728304617
9	0.009808823529	0.6497987609
10	0.011029411765	0.5176592954
12	0.013485294118	0.3823542712
13	0.014705882353	0.3116266450
14	0.015926470588	0.3173698729
15	0.017161764706	0.6457287568

Now we can fit the neural network to the training set. We will fit a non-linear network and use four hidden layers, each with five cells.

```
model = neuralnet(flux~date, data=train,
    hidden=c(5,5,5,5),
    linear.output=F)
```

Next, make a prediction **pre** using **compute**. The scaled value of the prediction will be in **pre$net.result**, so we scale it back to observed units by multiplying by the width and adding the minimum.

```
pre = compute(model,test[1])
predicted.flux = pre$net.result*widths[2]
    +minimums[2]  # column $net.result
```

We also scale the observed value, which we take from the test data, and compute the R^2 value.

```
observed.flux=test[,2]*widths[2]+minimums[2]
1-sum((observed.flux-predicted.flux)^2)/
    (sum((observed.flux-mean(observed.flux))^2))
```

0.872897091820931

Thus we get an $R^2 = 0.87$. This number will vary depending on the random values used in the initialization of the network, or by setting your random seed.

We can plot the network with

```
plot(model,rep="best",font=12,radius=0.1)
```

The network is illustrated in figure 8.8.

There is one input node and one output node. Some of the information does not fit in the output node as the visualization capability of the **plot.nn** function are limited.

To visually compare the results of the fit with the observed data we will overlay a scatter plot of the observed data with a line plot of the the predicted fit to the test data. The output is shown if figure 8.9.

```
# line plot (type="l") of predictions (scaled to true units)
#
plot(test$date*widths[1]+minimums[1],
    predicted.flux, type="l",
    xlab="Year",ylab="10.7 cm Solar Flux (10^-22 J/m^2)")

# overlay this with scatter plot of raw data
#
points(solarflux$date,solarflux$flux,cex=0.2,
    grid(lty="dotted",col="black")) # $ date column is year
```

Figure 8.8.: Neural network with 5 hidden layers, each with 5 nodes, that was fit to the solar flux data. The network took over 43,000 iterations to converge

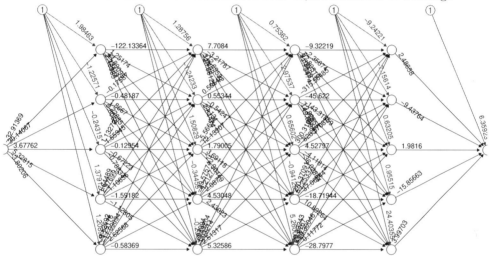

Figure 8.9.: Predictions based on test data (solid line) and all data (points). The network captures the 11 year cycle but it does not capture the more detailed structure (monthly) because the data is too sparse. There is only one point per month in the data file and more data is needed for that.

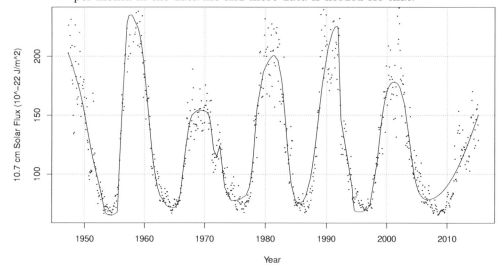

9. Regression Trees

Tree models can be used for both regression and classification. The algorithms for each are very similar. The acronym CART, for *Classification and Regression Trees*, is sometimes used.[1] When a tree is used to find a numerical result (e.g., a floating point value) it is called a regression tree. When one is used to place an item into a category such as "tastes like chicken" or "tastes like chocolate" it is called a classification tree. We will discuss regression trees in this chapter and return to classification trees in a later chapter.

In a **regression tree**, a problem is classified by repetitive subdivisions based on a decision process that minimizes the RSS error. If there are n explanatory inputs $x_1, \ldots, x_n$, then a linear regression problem fits a model of the form

$$y = a + b_1 x_1 + b_2 x_2 + \cdots + b_n x_n$$

In a regression tree, the domain is ultimately divided into K distinct, non-overlapping regions and a step function is fit. Let $\mathbf{x}$ be the vector of explanatory inputs. Then we are fitting

$$y = c_1 \delta(R_1, \mathbf{x}) + c_2 \delta(R_2, \mathbf{x}) + \cdots c_n \delta(R_K, \mathbf{x}) = \sum_{k=1}^{K} c_j \delta(R_k, \mathbf{x})$$

If there is only a single input then $\mathbf{x}$ collapses to the scalar x. Here δ is a set membership indicator function:

$$\delta(R_k, x) = \begin{cases} 1, & \text{if } x \in R_k \\ 0, & \text{otherwise} \end{cases}$$

Note that since x can only be a member of a single region, only a single term in the sum will be picked out by the sum. The argument to δ may be either scalar (for the single-dimensional input) or vector (for the multidimensional input).

Here's how the regression tree algorithm works. Suppose there are N training vectors and K regions. Let $y_1, \ldots, y_N$ be the corresponding values

---

[1] The acronym is generally attributed to the title of the book, Breiman, L, et. al. (1984) *Classification and Regression Trees.* Brooks/Cole.

of the response variable. The algorithm will continually seek to minimize the RSS error

$$\text{RSS} = \sum_{y_i \in R_1} (y_i - \hat{y}_1)^2 + \cdots + \sum_{y_i \in R_K} (y_i - \hat{y}_K)^2 = \sum_{j=1}^{N} \sum_{i=1}^{K} (y_j - \hat{y}_i)^2 \delta(R_i, x_j)$$

where $\hat{y}_i$ is the mean response in region i.

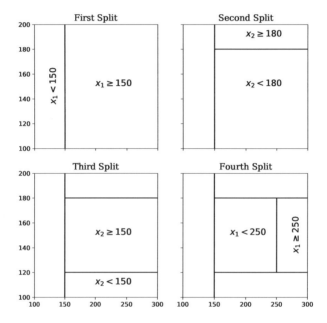

Figure 9.1.: Example of effect of cut points on explanatory variable domain for four successive decisions.

The problem is that we don't know what the K partitions are initially and we need to find them. We start with a full domain. Then we define **cut points** (figure 9.1) in each explanatory variable. If the explanatory variables are $x_1, x_2, \ldots, x_n$ then a cut point is a value z such that divides the dimension x_j into two regions, $x_j < z$ and $x_j > z$. We find cut points that minimize the RSS when the points are split over the two regions. Each variable is considered, so the split might be parallel to any axes. A region is split only along one explanatory variable at a time, so the regions remain rectangular. Once a cut point is found, the process is repeated again in each new sub-region.

At each division, we can predict the response (the response variable, y). This is the mean of the training observations in any given region. As we

Figure 9.2.: The response function is a multidimensional step function on the partitioned domain. The level of each plateau is the mean of the training values in that sub-domain.

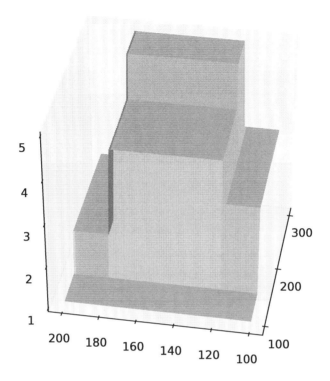

repeat this process then we get a hierarchy of values, hence a regression tree. The values of the response are plateaus over hyper-rectangular regions in the domain (figure 9.2. Iff the process is repeated too long there will be one leaf on the tree for each value in the training set, which will have been overfit! Thus the recursive subdivision is normally stopped at or before there are a minimum number of training points in each subdomain.

Example Using tree

We will do an example[2]. using the car data set from the UCI/ML repository (see page 57). First we read the data file and then edit out any rows that

---

[2]Full documentation is available at https://cran.r-project.org/web/packages/tree/tree.pdf

have missing data.

```
cars=read.table("./../datasets/cars.csv",
  TRUE, sep=",",na.strings="")
  okcars=na.omit(cars)
```

The first time we use **tree** we will need to install the package. Every time we run the package we will need to load the library. In jupyter we usually still need to specify the repository, even for many packages that are in the cran repository.

```
install.packages("tree", repos='http://cran.us.r-project.org')
library(tree)
```

We can create a regression tree with the **tree** function.

```
cars.tree.model = tree(mpg ~ displ+hp, data=okcars)
```

We want to plot the tree, but if we plot it on the screen it will be very TALL. So we reset the default screen plotting parameters.

```
library(repr)
options(repr.plot.width=7, repr.plot.height=4)
```

Then we plot the tree. The **plot** function will plot a skeleton of the tree, but it will not annotate it. To get text on it, we need to set the text parameters. We can use the default parameters with the **text** function.

```
plot(cars.tree.model)
text(cars.tree.model, cex=1.)
```

By default, text is 12 points. Using the **cex** parameter changes the text height by multiplying 12 by **cex**. The tree plotted is shown if figure 9.3.

There is a function **partition.tree** that will plot the partition in the domain if there are only one or two variables in the domain. First we will reset the screen window to 6 × 6. Sometimes it helps to visualize the information if we plot a projection of the data onto the plane being plotted. The **plot** and **partition.plot** must be in the same jupyter cell.

Figure 9.3.: The regression tree produced by **tree** on the cars data.

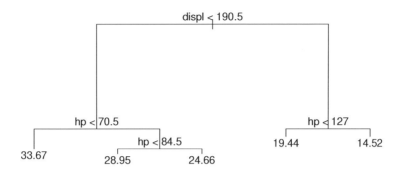

Figure 9.4.: The partition plot tree produced by **partition.tree** on the cars data. The small circles are a projection of the data onto the displacement/hp plane. The numbers are the miles per gallon values. These correspond to the plateau levels of the regression function.

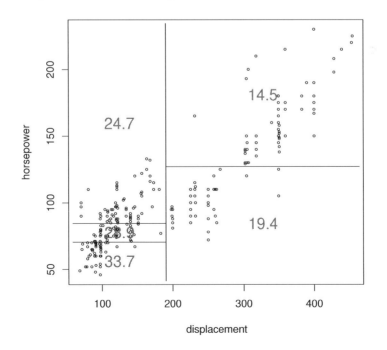

```
options(repr.plot.width=6, repr.plot.height=6)
plot(okcars$displ,
     okcars$hp,
     cex=.5, # this makes the little circles smaller
     ylab="horsepower",
     xlab="displacement")
partition.tree(cars.tree.model,
     ordvars=c("displ","hp"),
     add=TRUE,  # if this is not set to TRUE,
                # it will erase the first plot!
     cex=1.5,   # cex scales the font size
     col="red")
```

Example Using rpart

We will do an example[3]. using the car data set from the UCI/ML repository (see page 57). First we read the data file and then edit out any rows that have missing data.

```
cars=read.table("./../datasets/cars.csv",
  TRUE, sep=",",na.strings="")
  okcars=na.omit(cars)
```

Load the **rpart** library.

```
library{rpart}
```

We build the classification tree with **rpart**. There are two methods: (a) **"anova"** for regression; and (b) **"class"** for classification. We will return to the **"class"** method in chapter 18.

```
car.model = rpart(mpg ~ displ+hp, data=okcars,method="anova")
```

---

[3]Full documentation of **rpart** is available at https://cran.r-project.org/web/packages/rpart/rpart.pdf

Table 9.1. `rpart` options

`rpart(formula, data, weights, subset, (`options`))`
`rpart(data, weights, subset, `options`)`

Positional Parameter	Default Value	Description
`formula`	(none)	A formula or data frame. If it is a formula there may not be any interaction terms.
`data`	(none)	Required if formula is used. Data set, if formula is given.
`weights`	(none)	Optional weights to use for different case.

Parameter	Default	Description
`control`		Optional list of options to pass to `rpart.control`
`cost`		Option costs vector. Must be non-negative.
`method`	(*)	If missing, will try to guess based on data. Determines splitting criterion.

value	description
`"anova"`	Regression tree.
`"class"`	Classification tree. Uses Gini index or log likelihood.
`"exp"`	Survival tree.
`"poisson"`	Used to model rates. Uses a Poisson likelihood ratio.

Parameter	Default	Description
`model`	FALSE	Flag to keep a copy of the model in the results.
`na.action`	na.rpart	How to handle missing data. The default will remove all rows for which either the response (y) or all the predictors (x_i for all i) are missing.
`parms`		Additional parameters for method.
`x`	FALSE	Flag to keep a copy of the predictor data in the results.
`y`	FALSE	Flag to keep a copy of the y data in the results.

The model determines when to converge by calculating a complexity parameter `cp`. The default value is 0.01. This complexity parameter determines when further splitting will not significantly change the output of the model.[4] The value of `cp` that is used to determine convergence can be reset by the function `rpart.control` (table 9.2).

---

[4] See Therneau and Atkinson, *An Introduction to Recursive Partitioning Using the RPART Routines* https://cran.r-project.org/web/packages/rpart/vignettes/longintro.pdf

Table 9.2. **rpart.control** options

Parameter	Default	Description
cp	0.01	Complexity Parameter for convergence.
maxcompete	4	Maximum competitor splits saved.
maxdept	30	Tree depth.
maxsurrogate	5	Maximum surrogate splits. A surrogate split is a split assigned to a variable that is used when there is insufficient data for the primary splitting variable.
minbucket	(*)	Minimum number of data points in a leaf. (*) Will be set to **minsplit**/3 if not otherwise initialized.
minsplit	20	Minimum number of data points in a region or it will not be split.
xval	10	Number of cross validations, i.e., do 10-fold cross validation.

We can plot a history of the convergence of the complexity parameter with **cp** (see figure 9.5) and print a table of the successive values with **printcp**.

```
printcp(car.model)
plotcp(car.model)
```

```
Regression tree:
rpart(formula = mpg ~ displ + hp, data = okcars, method = "anova")

Variables actually used in tree construction:
[1] displ hp

Root node error: 23819/392 = 60.763
n= 392

          CP nsplit rel error  xerror     xstd
1 0.580331      0   1.00000  1.00306  0.061393
2 0.110608      1   0.41967  0.45053  0.038430
3 0.042452      2   0.30906  0.34934  0.030705
4 0.027348      3   0.26661  0.31026  0.029991
5 0.010000      4   0.23926  0.27929  0.026583
```

To plot the tree produce by **rpart** we can use the package[5] **rpart.plot**. This has two functions in it that are useful, a basic plotting function **rpart.plot** and a highly configurable plotting function **prp**.

---

[5]Full documentation of **rpart.plot** is available of https://cran.r-project.org/web/packages/rpart.plot/rpart.plot.pdf.

Figure 9.5.: Convergence of the complexity parameter using default settings for the car data.

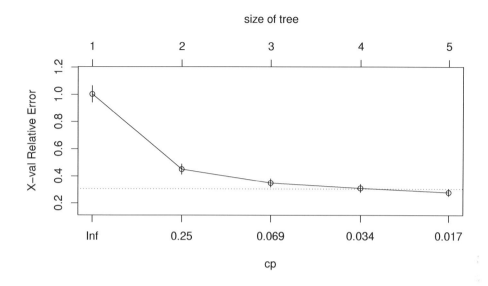

```
install.packages("rpart.plot",
    repos='http://cran.us.r-project.org')
library(rpart.plot)
```

To get the basic plot with the default options,

```
rpart.plot(car.model)
```

The output is shown in figure 9.6. To get the configurable plot with the default parameters, use

```
prp(car.model)
```

The output is shown in figure 9.7.

Making Predictions with rpart

Getting the regression tree wouldn't be very useful if we couldn't also make predictions automatically (coding the function manually would be very tedious). There is a function **predict** for this.

Figure 9.6.: Output of **rpart.plot(car.model)** using default values for all parameters.

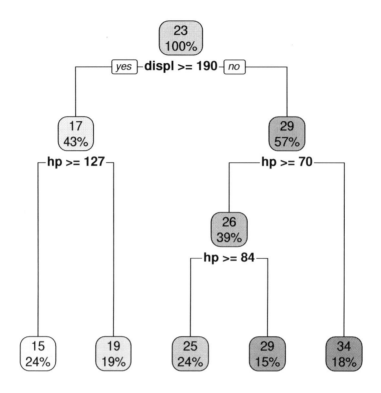

We illustrate predictions by generating a toy data set. We will produce a set of 100 data points that is randomly scattered about the line $y = 3x + 1$. The scatter is normal with mean zero and standard deviation 0.25. We generate and plot the data as follows.

```
xvals=seq(-1,1,length.out=100)
yvals = 1+3*xvals+rnorm(100,mean=0,sd=.25)
df=data.frame(x=xvals,y=yvals)
model=lm(y~x, data=df)
plot(xvals,yvals,xlab="x",ylab="y", pch=2)
abline(model)
```

The plot is illustrated in the top frame of figure 9.8.

First fit the toy model the same way we fit the car model.

```
library(rpart)
xy.model=rpart(y~x,df,method="anova")
```

Figure 9.7.: Output of `prp(car.model)` using default values for all parameters.

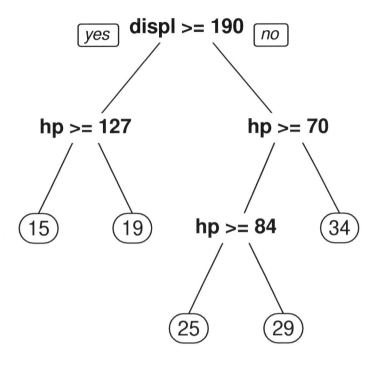

Then to produce a prediction, we generate a set of x-values for which we want to predict y values. We can use the **seq** function for this. To get a pretty plot, we will generate 1001 equally spaced points in the interval (-3, 1).

```
x=seq(-1,1,length.out=1001)
new.xvals=as.data.frame(x)
```

The predicted values are produced using **predict**.

```
y.predicted=predict(xy.model,new.xvals)
```

The plot is shown in the bottom frame of figure 9.8.

```
plot(x,y.predicted,type="l",
     xlab="x",ylab="y")
points(xvals,yvals, pch=2)
```

Figure 9.8.: Top: Data used for illustration of **predict** of regression tree fitted with **rpart**. The line is a least squares fit. Bottom: Step function produced by **rpart** using default parameters.

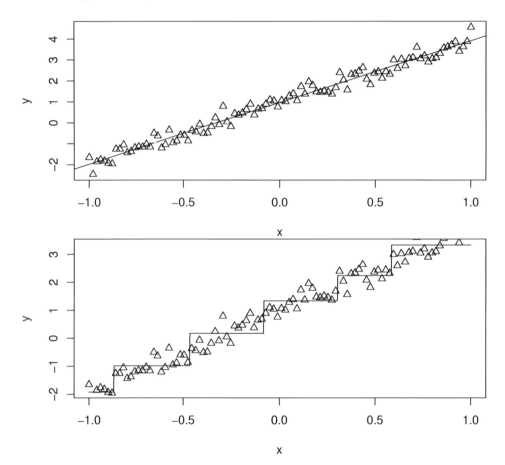

Part III.
Classification

Classification or categorization tasks involving placing items into their proper categories. They fall (broadly) into two types of problems:

- **Supervised Learning**. In supervised learning, we train a model with vectors that are marked with the correct answers. Feature vectors may contain information like bark color, leaf shape, tree height, wood hardness, etc. For example, if we have a data set that identifies the features of different types of trees, the vectors will be labeled as *elm*, *oak*, or *rowan*. A trained model will be given a vector that is unlabeled and then decide whether it is a *rowan tree*, and *elm tree*, or an *oak tree*.

- **Unsupervised Learning**. Rather than being told what type of tree each feature vector represents, the learning algorithm will contain the feature vectors, but not the the names of the trees. The algorithm will have to group similar trees together to figure out which trees fall into common categories. An trained model will say whether a new feature vectors falls into *category 1*, *category 2*, etc.

10. Logistic Regression

Logistic regression[1] is an example of a supervised classifier. It is easiest to visualize and describe when there are only two categories, but the method can be extended to multi-class categories.

In the simplest case there are two categories C_1 and C_0. We want to find a probability function $P(C_i|x)$ that will make this assignment. Let

$$y = P(C_1|x)$$

Then since there are only two categories,

$$1 - y = P(C_0|x)$$

The log of the odds ratio for C_1 is

$$z = \ln \frac{y}{1-y} = \text{logit}(y)$$

Solving for y gives

$$y = \frac{1}{1 + e^{-z}}$$

It can be shown that for normally distributed variables, z depends linearly on x.[2] Thus we have the logistic function

$$y = \frac{1}{1 + e^{-(ax+b)}}$$

where a and b are undetermined parameters.

If if there are multiple categories, the probability function becomes

$$y = \frac{e^{\mathbf{x}^T \mathbf{w}_k}}{\sum_k e^{\mathbf{x}^T \mathbf{w}_k}}$$

where the sum is over all categories (classes), $\mathbf{x}$ is a feature vector, and $\mathbf{w}_k$ is a vector of weights for class k.

---

[1] Cox, D. (1958) *The regression analysis of binary sequences.* Journal of the Royal Statistical Society B. **20**: 215-242.

[2] See E. Alpaydin, *Introduction to Machine Learning*, 2nd. ed., Chapter 10.5.

This is sometimes called the **softmax** function.

Logistic Regression is typically implemented using an iterative maximum likelihood estimator. A maximum likelihood estimator (MLE) finds the maximum value of the likelihood or log-likelihood function (natural logarithm of the likelihood), or the most probable outcome. Under certain conditions a MLE will produce the same result as least squares minimization. Typical iteration methods include variations on gradient descent algorithms, which attempt to follow the shortest path to an energy minimum (e.g., the minimum RSS error) by iterating on a formula similar to Newton's method.[3]

Logistic Regression in R

We will illustrate logistic regression in R using a dataset from the UCI machine learning repository. The data set is at `"https://archive.ics.uci.edu/ml/machine-learning-databases/adult/adult.data"`. Rather than actually downloading the file (which you can if you want) you should just copy the text of the filename into an R variable and call that variable whatever you like. Say we call it `url`. To load the data set directly over the internet,

```
url="FILENAME"
dataset=data.frame(read.csv(url, header = TRUE, sep = ","))
```

where `FILENAME` is replaced with the text of the URL. This data set contains 32,560 demographic records on adult Americans, with information about their educational level, race, employment, and income level. Here is an example of a typical entry (row 6 of the data set), along with their column headers.

```
print(dataset[6,])
```

```
    X39 State.gov X77516 Bachelors X13           Never.married
6    49   Private 160187       9th   5  Married-spouse-absent

      Adm.clerical Not.in.family White    Male X2174 X0 X40
6     Other-service Not-in-family Black  Female     0  0  16
```

---

[3] See chapter 3.3.2 of Y. Abu-Mostafa et. al., *Learning from Data* for a detailed discussion of the gradient descent algorithm for logistic regression.

```
   United.States X..50K
6       Jamaica  <=50K
```

The columns **Bachelors** and **X13**, for example give educational attainment as character strings and continuous levels.

```
sort(unique(dataset$X13))
```

```
1 2 3 4 5 6 7 8 9 10 11 12 13 14 15 16
```

```
sort(unique(dataset$Bachelors))
```

```
10th 11th 12th 1st-4th 5th-6th 7th-8th 9th Assoc-acdm
Assoc-voc Bachelors Doctorate HS-grad Masters Preschool
Prof-school Some-college
```

The column **X..50K** classifies income as above or below $50,000/year.

```
as.character(unique(dataset$X..50K)) #$ for column selection
```

```
' <=50K' ' >50K'
```

We will ask the following question from this data set: can income (above or below $50K) be predicted from level of educational attainment?

To answer this question we will only consider two columns in the data set: **X13** and **X..50K**. Lets first extract them to arrays, and then paste the arrays into a data frame.

```
edu=dataset$X13
income=dataset$X..50K
newdata=data.frame(edu,income)
print(head(newdata,5))
```

```
  edu income
1  13  <=50K
2   9  <=50K
3   7  <=50K
4  13  <=50K
5  14  <=50K
6   5  <=50K
7   9   >50K
8  14   >50K
```

Actually, we don't want to use that data frame! We want the data to be in 1's and 0's. Instead we'll build a new data frame that way. We'll revalue the flags to `" <=50K"` to `0` and `" >50K"` to `1` using the function `revalue` in the library `plyr`.

```
library{plyr}
P = revalue(nincome,(c(" <=50K"=0)))
Q = revalue(P,(c(" >50K"=1)))
R = as.numeric(as.character(Q))
newdata = data.frame(edu,R)
print(head(newdata,8))
```

```
  edu  R
1  13  0
2   9  0
3   7  0
4  13  0
5  14  0
6   5  0
7   9  1
8  14  1
```

In order to get an idea of the actual distribution of grade levels and income, let's build histograms. We'll extract the rows with **income** flag equal to 1 first, and put them into a data frame **hiincome**. Then we'll build a histogram of the **edu** column from the **hiincome** data frame. Next we'll extract the rows with **income** flag of 0 into a data frame **loincome**. Finally, we'll build a histogram of the **edu** column of the **loincome** data frame.

```
hiincome=newdata[newdata$income==1,]
options(repr.plot.width=8, repr.plot.height=4.5)
hist(hiincome$edu,main="Above $50,000/Year",
  xlab="Education", xlim=c(1,16))
axis(side=1,at=seq(1,16))
loincome=newdata[newdata$income==0,]
hist(loincome$edu, main="Below $50,000/Year",
  xlab="Education")
axis(side=1,at=seq(1,16))
```

An examination of these plots (shown in figure 10.1) shows that there is some distinction between the groups, but that there is great deal of overlap. Since there are individuals at all levels of educational attainment in both categories, the predictor variable **edu** is not likely to be a sufficient method of predicting which category a single adult will fall into. Nevertheless we will attempt to use it as a predictor variable for logistic regression and see where it takes us.

Figure 10.1.: Histograms of entire adult data set, broken up into two categories, as a function of educational attainment. The top shows individuals with income above $50K/year, and the bottom those with incomes below $50K/year.

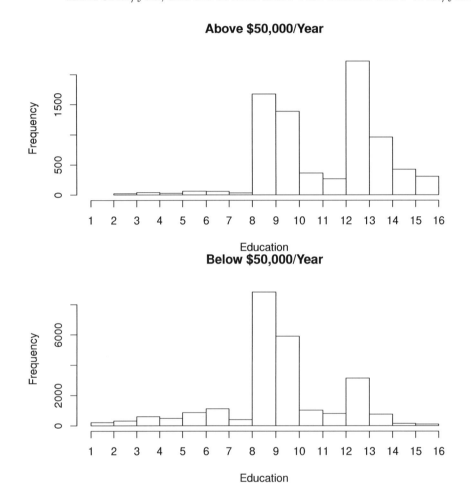

First, we will generate the test and training data sets.

```
nvals=nrow(newdata)
ind=sample(1:nvals, .75*nvals)

train=newdata[ind,]
test=newdata[-ind,]
```

Logistic regression is performed with **glm**

```
model=glm(income~edu, data=train, family=binomial)
print(model)
```

```
Call:  glm(formula = income ~ edu, family = binomial, data = train)

Coefficients:
(Intercept)          edu
   -5.0019       0.3628

Degrees of Freedom: 24419 Total (i.e. Null);   24418 Residual
Null Deviance:         27000
Residual Deviance: 24000           AIC: 24000
```

The model **Coefficients** tell us that a best fit probability is given by

$$P(y > 50K) = \frac{1}{1 + e^{-(a+bx)}} = \frac{1}{1 + e^{-(-5.00 + 0.36 \times \text{edu})}}$$

Although we *could* use this equation to make predictions, we don't have to – we can use the **predict** function. To generate a smooth curve for the plot, we'll generate points at intervals of 1/10 of a grade, even though those levels don't actually exist. Its easiest to create a special data frame with the sequence of values where we want to plot the predictions. Then when we call **predict**, we need to tell it to generate a **type="response"** because we are doing a probability function.

```
edu=seq(1,16,by=.1)
predictinput=data.frame(edu)
pre.model=predict(model,newdata=predictinput,
   type="response")
```

Here's how we could plot the observed (training) data and the predicted probability.

```
options(repr.plot.width=8, repr.plot.height=4.5)
plot(edu,pre.model,type="l",xlab="Years of Education",
    ylab="Probability", ylim=c(0,1))
points(train$edu, train$income)
grid()
```

Observe from figure 10.2 that the where the actual points have been plotted, the figure is not very informative (note that figure 10.2 has a few more

Figure 10.2.: Logistic regression on income as a function of educational attainment. The threshold of $P = .5$ is indicated with a horizontal line. The horizontal line intersects the logistic curve at $x = -a/b$ where a is the intercept and b the slope of the fit.

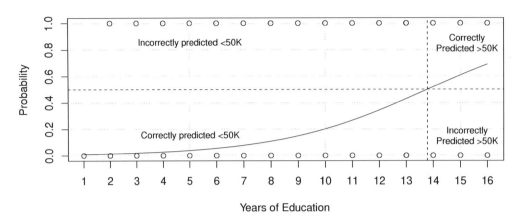

annotations that are not shown in the code). This is because when the data is plotted this way we can't actually see how many individuals are in each level (grades 1 through 16, e.g.). Second, from the figure, we observe that the probability never gets very strong. Usually we take a probability of 1/2 as the cutoff between categories. But there is never a very high likelihood of $P > 0.5$. So as expected, this may not be a very good model. Data points that fall in the upper right part or lower left of the plot are predicted correctly but the points that fall in the upper left or lower right are not predicted correctly. Finally, note that we never used the test data set. This is because we will use the test data set for evaluating the data. Because we want to say some things about evaluating binary classifiers that are more general than just logistic regression, we will continue with this example in chapter 11.

11. Evaluating Binary Classification

In supervised learning we have an absolute measure of when a classification is correct and when it is not. The terminology used by scientists can get pretty confusing here. Since we have two classes (or categories) we identify one of them as the category of interest. If something is classified as being in that category, we call it a **positive** classification. If something is classified as being the in the other category, we call it a **negative** categorization.

The definition of which category is positive and which category is negative is completely arbitrary and up to the experimenter (analyst) but must be made completely clear to begin with. For classification tasks that are performed with methods like logistic regression, we often say something like $y = 1$ for the positive class and $y = 0$ for the negative class, but again, this is completely arbitrary. We could just as easily have replaced the set $(0, 1)$ with (yes, no) or $(+, -)$, or (p, n) (for positive, negative). A **classification model**, such as the ones produced with `glm` for logistic regression, is any mapping from instances (domain or x values) to class values (y values).

For the remainder of this section, we will assume that the class values are zero (for a negative) and one (for a positive). When we create a model we provide a training set, or a collection of **exemplar** class values. We say that an exemplar class value is

- **positive** if $y = 1$.
- **negative** if $y = 0$.

The model will generate predictions y_p for class values. For logistic regression, each prediction is a continuous probability value, and the actual class is determined by whether or not the the probability is greater than or less than a cutoff value (usually taken as $1/2$). We say that a predicted class value is a

- **true positive** (TP) if $y_p = 1$ and $y = 1$.
- **false positive** (FP) if $y_p = 1$ and $y = 0$.
- **true negative** (TN) if $y_p = 0$ and $y = 0$.
- **false negative** (FP) if $y_p = 0$ and $y = 1$.

If a predicted value is a true positive or a true negative, the prediction is correct. If the predicted value is false positive or a false negative, the prediction is wrong.

Ideally we would like to build classifiers that minimized the number of false negatives and false positives. To visualize the success (or failure) of a classifier we count the number of times each item in the test set is predicted fall into one of these categories. We denote the numbers of true and false positives and negatives as TP, FP, TN, and FN. Suppose we denote the negative class by C_0 and the positive class by C_1 (you can remember the class names because the subscript in the same as the y-value). Then we can display these four quantities in a **confusion matrix** as follows.

		Actual Class	
		C_0	C_1
Predicted	C_0	TN (True Negative)	FN (False Negative)
Class	C_1	FP (False Positive)	TP (True Positive)

Observe that the column sums of the confusion matrix are the total numbers of negative and positive exemplars, respectively: Let

$$N_p = TP + FN = \text{total number of positive exemplars}$$
$$N_n = TN + FP = \text{total number of negative exemplars}$$

We can define the **true positive rate** (tpr), **recall**, (R), or **sensitivity** as

$$tpr = R = \frac{TP}{N_p} = \frac{TP}{TP + FN}$$

The **false positive rate** (fpr) or **fallout** (F) is defined as

$$fpr = F = \frac{FP}{N_n} = \frac{FP}{TN + FP}$$

Closely related to the fallout (and sometimes used in its place) is the **specificity**,

$$S = 1 - F = \frac{TN}{FP + TN}$$

The **precision** gives the ratio of true positives to total positives. It gives the proportion of values, among all those *predicted* to be in category C_1, which were predicted correctly.

$$P = \frac{TP}{TP + FP}$$

The **accuracy** of a model is defined as the proportion of items that were correctly predicted.

$$A = \frac{\text{Number of Correct Predictions}}{\text{Number of data points}} = \frac{TP + TN}{N_p + N_n}$$

The accuracy is the sum of the diagonal elements of the confusion matrix divided by the sum of all the elements in the matrix.

In the best of all worlds we would like a predictor with high precision, high recall (correct positive rate), and high accuracy (near 1). This is not always possible. Therefore a more balanced measure, the F1 score, sometimes used. The F_1 score is the harmonic mean of precision and recall;

$$F_1 = \frac{2PR}{P + R}$$

The reason for using a harmonic mean rather than a true average is illustrated by the following example:[1] Suppose the recall is high ($R \approx 1$) and the precision is low ($P \ll 1$). Then the arithmetic mean is around 0.5 while the harmonic mean is $\approx 2P$.

Confusion Matrix in R

We will continue the education versus income example begun in chapter 10 right where we left off. We need to load the **caret** library[2]

```
library(caret)
```

---

[1] From K. Murphy, *Machine Learning*.
[2] Kuhn, M, et. al. (2018). *caret: Classification and Regression Training* at https://cran.r-project.org/package=caret. In particular, there is a useful vignette at https://cran.r-project.org/web/packages/caret/vignettes/caret.html and a reference manual https://cran.r-project.org/web/packages/caret/caret.pdf.

Then we will make a prediction, this time using the test data set (last time we used the training set).

```
pre=predict(model,newdata=test,type="response")
```

The prediction are continuous numbers between zero and one so we need to convert them zero and one values. The expression **as.numeric(pre>0.5)** will return an array of numeric values (1's and 0's) by putting a threshold of 0.5 on the continuous data. However, the **confusionMatrix** function requires its inputs to be converted to factors. So we need to wrap this with **as.factor**.

```
confusionMatrix(
  data=as.factor(as.numeric(pre>.5)),
  reference=as.factor(test$income))  # $ for column
```

```
Confusion Matrix and Statistics

          Reference
Prediction    0    1
         0 5941 1511
         1  256  432

               Accuracy : 0.7829
                 95% CI : (0.7738, 0.7918)
    No Information Rate : 0.7613
    P-Value [Acc > NIR] : 2.011e-06

                  Kappa : 0.2326
 Mcnemar's Test P-Value : < 2.2e-16

            Sensitivity : 0.9587
            Specificity : 0.2223
         Pos Pred Value : 0.7972
         Neg Pred Value : 0.6279
             Prevalence : 0.7613
         Detection Rate : 0.7299
   Detection Prevalence : 0.9155
      Balanced Accuracy : 0.5905

       'Positive' Class : 0
```

ROC Curve

A commonly used tool is the **ROC curve**.[3] This is a plot of the true positive rate (i.e., recall) as a function of the false positive rate (i.e., fallout) at various threshold settings. The scale on each axis runs from zero to one. (Its axes are often mis-labeled as a plot of true positive vs false positive, but this is not correct, because without the normalizations (division by N_p and N_n, respectively) the values on the axes could not be between zero and one.) Sometimes the x axis is labeled with specificity rather than fallout; when this is done, the direction of the x axis is reversed, with higher values pointing to the left, so that the actual curve is unchanged.

Calculation of the ROC curve only depends on the TP rate and FP rate. A point in *ROC space* give values (tpr, fpr) and corresponds to a *discrete classification*, i.e., the result of a single class label. The classification must be varied through the space to get the curve that corresponds to a particular data set. These can be calculated from vectors of exemplar values (e.g., the y values from a test data set) and predicted data (e.g., the the $\hat{y}$ values that are obtained when the x values from a test set are passed into a model).

To see why the ROC curve traced out by the classification of a particular data set is useful, consider the significance of different locations in ROC space.

- The origin (0,0) corresponds to $tpr = fpr = 0$; while there are no false positives, there are no correct classifications.

- The point $(0, 1)$ corresponds to $tpr = 1$ and $fpr = 0$. All points were were identified correctly, and the data was perfectly classified.

- The point $(1, 1)$ corresponds to $tpr = fpr = 1$ corresponds to unconditional positive classification.

Generally, a point in ROC space is "better" if the true positive rate is better than than the false positive rate. This corresponds to the region

---

[3] ROC stands for Receiver Operating Characteristic or something silly like that. The name originated with radar science and has little meaning. Use of the ROC curve for machine learning is describe in detail in Fawcett, T. (2006) *An introduction to ROC analysis.* Pattern Recognition Letters. **27**: 861-874.

$tpr > fpr$ or the upper left hand triangle. A "good" classifier will have its ROC curve in the upper left, and a "bad" classifier will have its ROC curve in the lower right. The better the classifier the more its elbow will be pulled towards the upper left hand corner.

We can computer the ROC curve from basic principles in R. The function **my.roc** defined below, where **Examples** is an exemplar vector and **Pred** is a prediction vector does just this.[4]

```
my.roc=function(Examples,Pred){
    P=sum(Examples==1)
    N=sum(Examples==0)
    N.pts=P+N
    # sort data by pred
    temp=data.frame(Examples,Pred)
    temp=temp[order(-Pred),]
    #
    L=temp$Examples
    f=temp$Pred
    #
    FP=0; TP=0;
    R.x=rep(0,N.pts); R.y=rep(0,N.pts)
    f.previous=-99999 # - infinity
    j=1
    #
    for (i in 1:N.pts){
        if(f[i]!=f.previous){
            R.x[j]=FP/N
            R.y[j]=TP/P
            j=j+1
            f.previous=f[i]
        }
        if(L[i]>0.5){TP=TP+1}
        else{FP=FP+1}
    }
    R.x[j]=FP/N
    R.y[j]=TP/P
    fallout=head(R.x,j)
    recall=head(R.y,j)
    data.frame(fallout,recall)
}
```

This code is shown because it is often very confusing to students how it is possible to get a complete plot of true positive rate vs false positive rate

---

[4]This code is based on Algorithm 1 from Fawcett, *op.cit.*

when it seems like single classification should only produce one value each of recall and fallout.

An identical plot can be produced using the function **roc** in the library **pROC**.[5] (There are also many other libraries that implement ROC curves so if you find one you like, feel free to use it instead.)

All we need to run **roc** is an exemplar vector and a prediction vector. We can use the same prediction vector **pre** that we used above for the confusion matrix, and we can use column **income** direction from the test data frame.

```
r=roc(test$income,pre)
print(r)
```

```
Call:
roc.default(response = test$income, predictor = pre)

Data: pre in 6197 controls (test$income 0) < 1943 cases
   (test$income 1).
Area under the curve: 0.7206
```

Since **roc** returns **specificity** and **sensitivity**, we will turn these into recall and fallout for plotting.

```
fallout=1-r$spec
recall=r$sens
plot(fallout,recall,type="l")
lines(0:1,0:1,lty="dotted")
abline(h=seq(0,1,by=.2), v=seq(0,1,by=.2), lwd=.2)
```

The plot is illustrated in figure 11.1.

Finally, we observe that **roc** also returned an AUC value (see the **print** output, above). The acronym AUC means *area under the curve*. The AUC is often used as a measure of the goodness of prediction. The closer this value is to one, the better the prediction. For comparison, the area under a 45-degree diagonal is 0.5. The AUC can be found by integrating the area

---

[5]Robin, X (2001) *pROC: an open-source package for R and S+ to analyze and compare ROC curves.* BMC Bioinformatics. **12**:77. The library is available at https://cran.r-project.org/package=pROC.

Figure 11.1.: ROC curve for income. Predictions using test data.

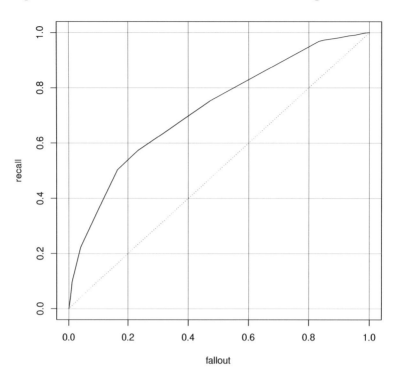

under the ROC curve. If you generate the ROC curve with **my.roc** then you can calculate the AUC with a simple trapezoidal rule integration.

```
my.AUC=function(roc){
    x=roc$fallout
    y=roc$recall
    AUC=0
    for (i in 2:nrow(roc)){
        AUC = AUC + 0.5*abs(x[i]-x[i-1])*(y[i]+y[i-1])}
    AUC
}
```

We can get the same results as those shown here with the following code (output not shown).

```
r=my.roc(test$income, pre)
print(my.AUC(r))
plot(r$fallout, r$recall, type="l")
```

12. Deep Learning - Classification with Keras

Keras[1] is a library for modeling neural networks. In this chapter we will illustrate a classification problem, that of identifying the written characters from 0 through 9.

Keras is primarily used as a Python package, and it includes several other popular libraries, notably TensorFlow. There is a wrapper for Keras that allows you to import it directly into R and build and evaluated neural network models.

Figure 12.1.: The first fifty characters from the training set of the MNIST data set.

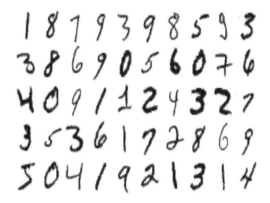

The data set we will use is the modified National Institute of Standards and Technology (MNIST) character set for numbers.[2] Keras includes a copy of the data set in its library, so once you install Keras you won't have to go out to the web page to download the data. However, the data is stored in a pretty unusual format, and if you want to really understand it, you may want to read the data file description on the MNIST web page. The data consists of a training set and a test set. The training set has 60,000 images and the test set has 10,000 images. Each image is a 28 by 28 pixel representation of a picture of one hand drawn number, such as the numbers drawn in figure 12.1.

---

[1] See https://keras.io/fordetails. Much of the material in this chapter is based on one of the Keras tutorials for beginners.
[2] Available at http://yann.lecun.com/exdb/mnist/.

First you have to install Keras. A normal installation will look something like this.

```
install.packages("keras")
library(keras)
install_keras()
```

```
Creating virtualenv for TensorFlow at
 ~/.virtualenvs/r-tensorflow
Upgrading pip ...
Upgrading wheel ...
Upgrading setuptools ...
Installing TensorFlow ...

Installation complete.
```

Most likely you will not have a normal installation. The most common thing that can go wrong is a missing prerequisite file. Keras will print out a message telling you what is missing. For example, on a linux operating system, you could get the following message:

```
Error: Prerequisites for installing TensorFlow not available.

Execute the following at a terminal to install the
prerequisites:

    sudo apt-get install python-pip python-virtualenv
```

This tells you what is missing and how to install it. After installing the missing software, go back and start from where the installation stopped the last time.

In the future all we should have to type into our notebook is

```
library(keras)
```

This will get us started running Keras models.

Let's start by loading the MNIST data set. Its built-in so its easy to get.

```
MNIST=dataset_mnist()
```

The data is stored as weirdly formatted arrays. To see this,

```
summary(MNIST)
```

```
       Length Class  Mode
train  2      -none- list
test   2      -none- list
```

So **MNIST** has two parts: **MNIST$train** and **MNIST$test**. Lets look at the training set.

```
summary(MNIST$train) # $ indicates part of a class
```

```
  Length   Class  Mode
x 47040000 -none- numeric
y    60000 -none- numeric
```

So **MNIST$train$x** has 47,040,000 items, and **MNIST$train$y** has 60,000 items. To see how they are shaped, we look at the dimension.

```
dim(MNIST$train$x)
```

60000 28 28

So we have a 3-dimensional array that is $60,000 \times 28 \times 28$. To turn this into something useful for classification, we need to turn each of the 28×28 sub-arrays into a $28 \times 28 = 784$ item long vector. The images in the raw data have values in a range of 0 to 255 (this is typical for an 8-bit gray scale image). We will convert this to a range of 0 to 1 as we did in our analysis of backpropagation networks.

We can do the reshaping with **array_reshape**. The second argument specifies the desired reshaping. We will make 60,000 vectors, each with length 784. We divide the whole thing by 255 to do the normalization.

```
x.train = array_reshape(MNIST$train$x, c(60000,784))/255
x.test  = array_reshape(MNIST$test$x,  c(10000,784))/255
```

One thing that will help us be able to visualize the data is to try to plot the figure. We can do this by converting each the images to rasters and using a raster plot. A raster is a reprsentation of an image as a matrix of pixels. In R we can use the function **rasterImage**. We can place an image on an already created plot by specifying the coordinate of the four corners of the raster.

The following function will plot a range of images starting with character **nstart** and stopping with character **nstop** in the test set, with 10 digits on a line. The digits are printed 10 to a line, with succeeding lines printer higher in the y-direction.

```
plotrows=function(data, nstart, nstop){
    n=nstop-nstart+1
    nrows=(n %/% 10) + 1       # total rows of images
    width = 10*28              # total pixels width
    height = nrows*28          # total pixels height
    plot(c(0,width), c(0,height),
        xlab="",ylab=" ",axes=FALSE)
    options(repr.plot.width=width/28,
        repr.plot.height=height/14) # make a nice aspect
    for (j in nstart:nstop){
        the.row = (j-1) %/% 10    # integer div, 10/row
        bottom = 28*the.row       # 28 pixels per row
        the.column = (j-1) %% 10  # remainder after division
        left = 28*the.column      # 28 pixels per column

        # convert to black on white instead of white/black
        im = 1-data[j,]           # extract image  from data

        dim(im)=c(28,28)          # convert to square array
        im = t(im)                # transpose, pretty print
        rasterImage(im,xleft=left,
                    ybottom=bottom,
                    xright=left+28,
                    ytop=bottom+28)
        }
}
plotrows(x.train, 1,50) # print training data 1 through 50
```

The last line, `plotrows(x.train, 1,50)`, will produce the image shown in figure 12.1.

Now lets look at the y data values – these give us the labels that tell us what the numbers in the figures really mean. We can just pull off a few at the top to look at them. Here are the labels of the first 25 values in the training set.

```
head(MNIST$train$y,25)
```

5 0 4 1 9 2 1 3 1 4 3 5 3 6 1 7 2 8 6 9 4 0 9 1 1

These are the numbers that label the first 25 images in figure 12.1 (reading left-to-right, bottom-to-top).

Before we do the fit, we want to convert the y-data using a **one-hot** encoding. In a one-hot encoding, each categorical variable is encoded as a binary vector consisting of all zeros and a single one (hence one hot bit and the rest are cold). Each categorical variable is encoded in such a way that no two variables have the same encoding. For example, here is a one-hot encoding of a categorical variable `Beatle` that can take on four possible values.

Value	One-Hot Encoding
"John"	$(0, 0, 0, 1)$
"Paul"	$(0, 0, 1, 0)$
"George"	$(0, 1, 0, 0)$
"Ringo"	$(1, 0, 0, 0)$

All of the encodings are orthogonal to one another, and all have length one. Using a one-hot encoding allows the model to find natural orderings in the data without being encumbered by artificial orderings or biases that might be induced by using a numerical scheme such as John = 1, Paul = 2, George = 3, Ring = 4.

In R, the function `to_categorical(variable, bits)` will produce a one-hot encoding of `variable`. If the number of `bits` specified is smaller than the number of classes (categories0 in the data set, an error will occur. If the number of bits is larger, the extra bits will be zero filled. If you know how many categories are in your data in advance this is ideal, so that your encoding won't waste extra space.

Since there are exactly 10 categories (the digits 0 through 9) in our data set we will ask for a 10-bit one-hot encoding.

```
y.train = to_categorical(MNIST$train$y,10)
y.test = to_categorical(MNIST$test$y,10)
```

This is what the first five encodings look like. Recall from above that the first five training vectors were pictures of the numerals 5, 0, 4, 1, 9.

```
head(y.train,5)
```

```
0 0 0 0 0 1 0 0 0 0
1 0 0 0 0 0 0 0 0 0
0 0 0 0 1 0 0 0 0 0
```

Ch. 12. Deep Learning with Keras

```
0 1 0 0 0 0 0 0 0 0
0 0 0 0 0 0 0 0 0 1
```

Table 12.1. Keras **layer_dense** Parameters

Parameter	Description
`activation`	Activation function. The default is linear, output = input. See table 12.2.
`activity_regularizer`	Regularizer that can be applied to the output. See https://keras.io/regularizers/ and note 2.
`batch_size`	Integer, how many pieces of information you expect the network layer to process at a time.
`bias_regularizer`	Regularizer that can be applied to the bias. See https://keras.io/regularizers/ and note 2.
`bias_constraint`	Constraint function for bias. See https://keras.io/constraints/ and note 1.
`dtype`	Data type of input.
`input_shape`	Number of inputs. Required if this is the first layer.
`kernel_initializer`	Function to use to randomly set initial weights. (see https://keras.io/initializers/).
`kernel_regularizer`	Regularizer for weight matrix. See https://keras.io/regularizers/ and note 2.
`kernel_constraint`	Constraint function for weights. See https://keras.io/constraints/ and note 1.
`name`	Optional (unique) name.
`trainable`	TRUE/FALSE, are your weights fixed or trainable.
`units`	Integer number of outputs from Layer.
`use_bias`	TRUE/FALSE, is there a bias vector.
`weights`	Initial weights matrix.

Notes. [1]Typical constraints are non-negativity, unit norm, and min/max norm.
[2]Regulators apply additional penalties during computation of the loss function.

We see that 0 is encoded by (1,0,...,0); 1 is encoded by (0,1,0,...); up through 9, which is encoded by (0,0,...,0,1), in this scheme.

Now that we have both the test and the training set, we need to build a model. A Keras Model consists of a sequence of layers of neurons. Each layer is built with a function **layer**_*sometype*. In the first layer, you need to specify the input shape, that is the number of input vectors.

To prevent over-fitting (modeling the noise rather than the actual data) we want to randomly select only a fraction of the training data during each iteration to work with. We do this by interspersing the dense layers with dropout layers.

Here is our model:

```
model = keras_model_sequential()
model %>%
  layer_dense(units = 256, activation = 'relu',
              input_shape = c(784)) %>%
  layer_dropout(rate = 0.4) %>%
  layer_dense(units = 128, activation = 'relu') %>%
  layer_dropout(rate = 0.3) %>%
  layer_dense(units = 10, activation = 'softmax')
```

You can use **summary** to print out a summary of the model once you have built it.

```
summary{model}
```

Table 12.2. Activation Functions in Keras Layers

elu	elu(x, alpha) Exponential linear. $y = x$ if $x \geqslant 00$ and $y = \alpha(e^x - 1)$ if $x < 0$.
linear	Identity function.
relu	relu(x, alpha, max_value) Rectified Linear. Returns x if $x \geqslant 0$; αx if $x < 0$. Capped at **max_value** if specified.
selu	selu(x, alpha, scale) Scaled Exponential Linear. equivalent to scale * elu(x, alpha).
sigmoid	Implementation depends on backend being used. If you are using TensorFlow, this will return $1/(1 + e^{-x})$.
softmax	Implementation depends on backend being used. If you are using Tensor Flow, this will return $e^{x_j \cdot W} / \sum_i e^{x_i W}$ where **W** is a weight matrix.
softplus	Returns $\log(e^x + 1)$.
softsign	Returns $x/(|x| + 1)$.
tanh	Hyperbolic tangent.

Ch. 12. Deep Learning with Keras

Table 12.3. Dropout Layers in Keras Models.	
`batch_size`	Use a fixed number of training vectors during each iteration.
`name`	Optional layer name.
`rate`	Fraction of the input data to drop during each iteration. Must be a floating point number between 0 and 1.
`seed`	Optional integer to be used as seed for random number generation.
`trainable`	TRUE/FALSE, can the weights be trained.
`weights`	Initial values for weight matrix.

This is the output of **summary**:

```
Layer (type)              Output Shape              Param #
=================================================================
dense_4 (Dense)           (None, 256)               200960
_____
dropout_3 (Dropout)       (None, 256)               0
_____
dense_5 (Dense)           (None, 128)               32896
_____
dropout_4 (Dropout)       (None, 128)               0
_____
dense_6 (Dense)           (None, 10)                1290
=================================================================
Total params: 235,146
Trainable params: 235,146
Non-trainable params: 0
```

After we build the model, we need to compile it. Compilation requires (a) an optimizer; (b) a loss function; and (c) a metric. The optimizer is the function that solves for the weight matrix (table 12.4). It does this by minimizing the loss function. The loss function used for categorical comparisons is the categorical cross entropy function. Let **p** be the target vector **q** be two predicted state vectors. Then

$$H(\mathbf{p}, \mathbf{q}) = -\sum_i p_i \log q_i$$

If **p** has a one-hot encoding, with only $p_k \neq 0$, then

$$H(\mathbf{p}, \mathbf{q}) = -\log q_k$$

The metric tells the optimize how to measure its performance after each iteration.

```
compile(model,
  loss = 'categorical_crossentropy',
  optimizer = optimizer_rmsprop(),
  metrics = c('accuracy')
)
```

Table 12.4. Optimizers in Keras[1]

`Adadelta`	A modification of Adadelta.[3]
`Adagrad`	An adaptive sub-gradient algorithm that uses different step sizes for different features. It is a modified version of stochastic gradient descent.
`Adam`	Adaptive Moment Estimation. A variation of RMSProp that includes second moments of the gradients in the averaging.[4]
`Adamax`	Varation of Adam that uses the infinity norm.
`Nadam`	Variation of Adam with Nesterov momentum.[2] There is an additional term in the momentum that may increase the convergence rate.
`RMSprop`	Root Mean Square Propagation. A gradient descent algorithm that divides the learning rate by the running average of the magnitudes of recent gradients.
`SGD`	Stochastic Gradient Descent

Notes: [1]For details see https://keras.io/optimizers/.
[2]See T. Dozat, "Incorporating Nesterov Momentum in Adam," http://cs229.stanford.edu/proj2015/054_report.pdf for a comparison of gradient descent algorithms.
[3]See Zeiler, "ADADELTA: An Adaptive Learning Rate Method," https://arxiv.org/abs/1212.5701 for details.
[4]See Kingma and Ba, "Adam: A method for stochastic optimization," https://arxiv.org/abs/1412.6980v8 for details.

Finally, we are ready to train the model. We train with the **fit** function in keras. We will do 30 iterations with a batch size of 128 and hold back 20% of each run for validation.

```
results=fit(model,
   x.train,
   y.train,
   epochs = 30,
   batch_size = 128,
   validation_split = 0.2)
summary(results)
```

```
        Length Class  Mode
params  8      -none- list
metrics 4      -none- list
```

We can look at the values of the individual parameters and metrics with the **$** operator.

```
results$params  # use $ operator to look at values
```

```
$acc
0.870791666666667 0.9408125 0.954229166666667 0.962125
0.965041666666667 0.9679375 0.972041666666667 0.973395833333333
0.974979166666667 0.976604166666667 0.9770625 0.978041666666667
0.97925 0.9805 0.981958333333333 0.9805625 0.982458333333333
0.9829375 0.983229166666667 0.983395833333333 0.983708333333333
0.985145833333333 0.984666666666667 0.985395833333333|
0.984791666666667 0.9846875 0.9863125 0.986333333333333
0.985604166666667 0.985791666666667
$loss
0.427643926262856 0.198583953003089 0.154141860355933
0.129832664340734 0.116330171391368 0.10582877390335
0.0954835125754277 0.0910539427498976 0.0847323289935788
0.0818170747881134 0.0766599763284127 0.0777039298713207
0.0716478073733548 0.0693771734281133 0.0645934438264618
0.0665825072688361 0.063620336268718 0.0590802522543818
0.0590776746788373 0.0600303629251818 0.0586656069848686
0.0542955303303897 0.0540380507353693 0.0530945444790026
0.0534871724257246 0.0553399253676956 0.0515446206314179
0.0543929766930329 0.0537154584531672 0.0532965152993177
$val_acc
0.953500000158946 0.965166666825612 0.970166666825612
0.970666666507721 0.975166666825612 0.973916666507721
0.976583333492279 0.977916666507721 0.977166666507721
0.977416666507721 0.976749999841054 0.979583333492279
0.977249999841054 0.978749999841054 0.977999999841054
0.978749999841054 0.980166666507721 0.978916666825612
0.980083333174388 0.980249999841054 0.979083333174388
0.979499999841054 0.979416666507721 0.980583333174388
0.981083333174388 0.980249999841054 0.981166666507721
0.979083333174388 0.979833333174388 0.980916666825612
$val_loss
0.160489712436994 0.12325706688563 0.105186254024506
0.0983676138495406 0.0904715795094768 0.0938584078500668
0.0886932785368214 0.0871986801947157 0.0890492789782584
0.0907606691693266 0.0959086238259139 0.0897049702337633
```

```
0.0946040730550885  0.0907590638779414  0.0980892686600952
0.0936858934223031  0.0965870961875577  0.0981971457696054
0.101240606700808   0.095308084258344   0.0973622127282821
0.0988708641687408  0.102880811558542   0.103930320226912
0.10097289773752    0.102810310442248   0.10340744528301
0.10757599282806    0.107583380208293   0.106028270952762
```

The output is compatible with **plot**.

plot(results}

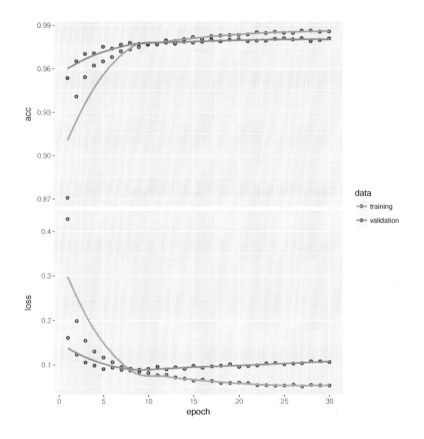

We can measure the accuracy and loss when we predict against the test data.

```
evaluate(model, x.test, y.test)
```

```
$loss
0.100639294248619
$acc
0.982
```

Ch. 12. Deep Learning with Keras

Finally, we can list out the predicted classes of the of the test data. Since there are 10,000 images in the test set we will on list the first ten of them here using the **head** function.

```
head(predict_classes(model, x.test),10)
```

    7 2 1 0 4 1 4 9 6 9

If we invoke our function **plotrows(x.test,1,10)** it will produce the image shown in figure 12.2. A close comparwison indicates that the network was correct nine out of ten times, misidentifying the 5 as a 6 in the second picture from the right.

Figure 12.2.: The first ten images in the test set.

13. K-Nearest Neighbors

K-nearest neighbors (KNN) is a supervised classification method in which cells are categorized according to their nearness. There is one input parameter, K, the number of near-neighbors to use. Given any test point **x**, KNN will look at the K training points that are closest to **x**, and it will examine which cluster each training point is labeled by. The method then calculates the conditional probability that **x** lies in each possible class. This is given by the fraction of training points that were in that class. Then the point is assigned to the most likely class. When $K = 1$ this reduces to a Voronoi tesselation (see page 181 and figure 22.1).

A knn classifier is implemented in the R package **class**.[1] The basic format of **knn** is

```
knn(train, test, cl, k = 1)
```

where **train** and **test** are matrices or data frames of training and test data; **cl** is a **factor** of class labels for the training set; and **k** is the number of classes to use. Other parameters are described in the detailed reference manual.[2]

We will demonstrate the use **knn** by building a toy model with two classes using the multivariate Gaussian **mvrnorm** in **MASS**. The multivariate Gaussian requires a set of means and a covariance matrix. We will use a simple diagonal covariance.

Load the packages first.
```
library(MASS)
library(class)
```

Generate two slightly overlapping clouds to use for training and one test could that overlaps them both. We'll set a random seed to make the results

---

[1] Ripley, B. (2005) *class: Functions for Classification*, which implements additional functions described in Venables, W. N. and Ripley, B. D. (2002) *Modern Applied Statistics with S. Fourth Edition.* Springer. The package, including full documentation, is available online at https://cran.r-project.org/package=class.

[2] Ripley, B., and Venable, W. (2015) *Package class*, available as https://cran.r-project.org/web/packages/class/class.pdf.

easily reproducible.

```
set.seed(25)
cloud1=mvrnorm(n=12,mu=c(2.2,2.4),Sigma=diag(c(.4,.2)))
cloud2=mvrnorm(n=12,mu=c(1.5,1.75),Sigma=diag(c(.2,.2)))
test=mvrnorm(n=15,mu=c(2,2), Sigma=diag(c(.4,.4)))
```

Here is the R to make a plot of the training data. We'll omit the test data from the plot for now (see figure 13.1).

```
plot(cloud1, col="red",xlim=c(0,4),ylim=c(0,4), pch=2)
points(cloud2, col="green", pch=6)
legend(.5, 3.5, c("class 1" ,"class 2"), pch=c(2,6),
    col=c("red", "green"))
```

Figure 13.1.: The training set used for **knn** consists of two labeled clouds of data.

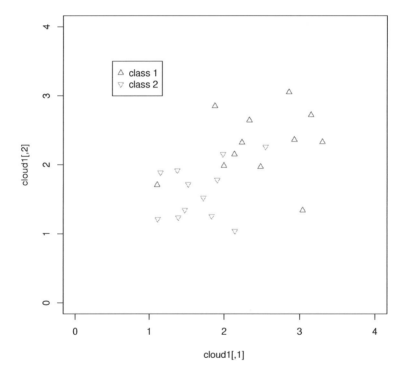

Before we can perform classification we need to (a) combine all the training examples together; and (b) we need to provide labels for the classes. We'll bind the examples together using **rbind**:

```
train=rbind(cloud1,cloud2)
```

We'll label the data in cloud 1 as class number 1, and the data in cloud 2 as class 2. We'll bind the two vectors together using **rbind** and then convert the whole thing to a `factor`.

```
cloud1.labels=rep(1,nrow(cloud1))
cloud2.labels=rep(2,nrow(cloud2))
train.labels=as.factor(c(cloud1.labels, cloud2.labels))
print(train.labels)
```

```
 [1] 1 1 1 1 1 1 1 1 1 1 1 2 2 2 2 2 2 2 2 2 2 2
Levels: 1 2
```

Now we are ready to classify. The function **knn** will return a vector of cluster numbers as a factors.

```
test.results=knn(train, test, train.labels)
print(test.results}
```

```
 [1] 1 1 1 2 2 2 1 2 1 2 1 2 2 1 1
Levels: 1 2
```

We can visualize the results by plotting them, sa shown in figure 13.2.

```
plot(cloud1, col="red",xlim=c(0,4),ylim=c(0,4),
     cex=.75, pch=2, xlab="x", ylab="y")
points(cloud2, col="green", cex=.75, pch=6)
legend(.5, 4, c("class 1 (training)" , "class 1 (predicted)",
               "class 2 (training)", "class 2 (predicted)"),
       pch=c(2,3, 6,4), col=c("red","red","green", "green"))
ntest=nrow(test)
for (j in 1:ntest){
    x=test[j,1]
    y=test[j,2]
    cluster=as.numeric(as.character(test.results[j]))
    if (cluster==1){points(x,y,pch=3,col="red")}
    else {points(x,y,pch=4,col="green")
}
```

Figure 13.2.: The results of classification by **knn** using **k=1** (top) and **k=3** (bottom) are shown here. The training set used for **knn** consists of two labeled clouds of data.

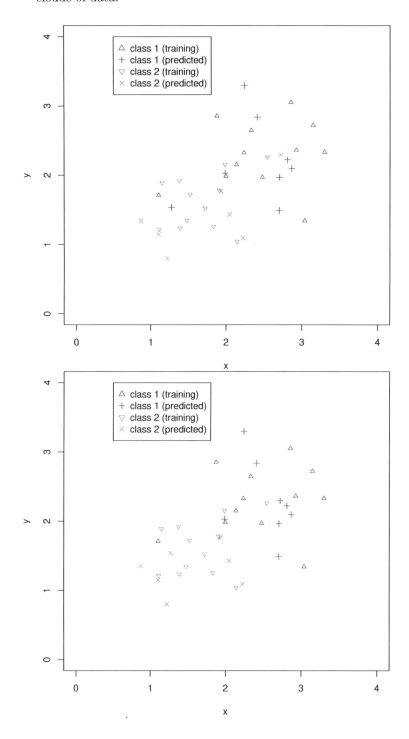

14. Naive Bayes' Classifiers

A Naive Bayes Clasifier use Bayes' Rule for classifying data. Bayes' Rules states that the probability of an observation **x** falling into a cluster k is

$$P(k|\mathbf{x}) = \frac{P(k)P(\mathbf{x}|k)}{P(\mathbf{x})}$$

where

$P(k) =$ Probability of cluster k
$P(\mathbf{x}) =$ Probability of observation **x**
$P(\mathbf{x}|k) =$ Probability of observation **x** assuming cluster k

If all of the features are independent (hence the *naive* part) then it is possible to write

$$P(\mathbf{x}|k) = \prod_{j=1}^{d} P(x_j|k)$$

where d is the number of features (the dimension of **x**). In general, this is a pretty big and not necessarily very accurate assumption. Consider what it means for the `auto-mpg` data set (page 57): in predicting `mpg`, we would assume that the factors `cyl`, `displ`, `hp`, `weight`, and `accel` are assumed to be independent. In fact, these factors are hardly independent. Cars with more cylinders and larger displacement tend to have higher acceleration, for example. So while naive Bayes' method is often a first method of choice by some analysts, care should be taken to ensure that the factors are independent first. One way to prepare the data might be to perform a dimensionality reduction, e.g., by PCA or factor analysis first.

There are a number of different implementations of Naive Bayes' classifiers in R. We will demonstrate with one that is very easy to use, in the `e1071` package.[1] We can load the library with

---

[1] The name E1071 was the mail code for the Department of Statistics and Probability Theory at the Technical University of Vienna, Austria, so the authors essentially name their software package after themselves. Meyer, D. et. al. (2017) e1071: Misc Functions of the Department of Statistics, Probability Theory Group (Formerly: E1071), TU Wien. Available at the CRAN URL, https://cran.r-project.org/package=e1071. Package documentation available as https://cran.r-project.org/web/packages/e1071/e1071.pdf.

```
library{e1071}
```

This example will use the Titanic data set that is one of the standard data sets available in R, and is based on the presentation given by Zhang.[2] The Titanic data set contains information about the individuals who survived the sinking of the titanic, broken down by gender, cabin class, and whether they were a passenger or crew member.

The data as stored in the Titanic data file is a rather complex table and rather difficult to visualize. It stores aggregate totals. We can load the data file and display part of it as a data frame as follows,

```
data(Titanic)
t.df= as.data.frame(Titanic)
head(t.df)
```

Class	Sex	Age	Survived	Freq
1st	Male	Child	No	0
2nd	Male	Child	No	0
3rd	Male	Child	No	35
Crew	Male	Child	No	0
1st	Female	Child	No	0
2nd	Female	Child	No	0

The data is summarized in table 14.1.[3] This is what the raw data looks like.

```
print(Titanic}
```

```
, , Age = Child, Survived = No

      Sex
Class  Male Female
  1st    0    0
  2nd    0    0
  3rd   35   17
  Crew   0    0

, , Age = Adult, Survived = No
```

---

[2] Zhang, Z (2016) *Naive Bayes Classification in R* Annals of Translational Medicine. 4:241.

[3] Table 14.1 was generated by manually cutting and pasting the data from R into a Libre-office spreadsheet.

```
          Sex
Class  Male Female
  1st   118      4
  2nd   154     13
  3rd   387     89
  Crew  670      3

, , Age = Child, Survived = Yes

          Sex
Class  Male Female
  1st     5      1
  2nd    11     13
  3rd    13     14
  Crew    0      0

, , Age = Adult, Survived = Yes

          Sex
Class  Male Female
  1st    57    140
  2nd    14     80
  3rd    75     76
  Crew  192     20
```

Table 14.1. Summary of Titanic Data					
Class	Gender	Age	Survived	Did Not Survive
1st	Male	Child	5	0
2nd	Male	Child	11	0
3rd	Male	Child	13	35
Crew	Male	Child	0	0
1st	Female	Child	1	0
2nd	Female	Child	13	0
3rd	Female	Child	14	17
Crew	Female	Child	0	0
1st	Male	Adult	57	118
2nd	Male	Adult	14	154
3rd	Male	Adult	75	387
Crew	Male	Adult	192	670
1st	Female	Adult	140	4
2nd	Female	Adult	80	13
3rd	Female	Adult	76	89
Crew	Female	Adult	20	3

Unfortunately the Titanic data is aggregated, and we want one data point for each person. This means that we have to unwrap the data. For example, the 35 male boys (male children) who did not survive need to be expanded into 35 data points. We can use Zhangs's **countToCases** function for this, which we modify into our function **as.cases**. The input to **as.cases** is a data frame **x** and a column name **C**. For every row in **x**, the value of **x[row, C]** is examined. Then that row is repeated that exact number of times. A new data frame is produced with all of the repeats.

```
as.cases = function(x, C) {
    indices = rep.int(seq_len(nrow(x)), x[[C]])
    x[[C]] = NULL # remove frequency column
    x[indices, ] # Extract the rows
}
```

To understand why this works we need to understand the **rep.int** function. It can take as input to vectors and then matches them to give repeats. For example, the following matches the 3 with the 2 and the 5 with the 4 to produce a vector with 3 repeated twice followed by 5 repeated 4 times.

```
rep.int(c(3,5),c(2,4))
```

3 3 5 5 5 5

Now consider what happens if we expand

```
seq_len(nrow(t.df))
```

1 2 3 4 5 6 7 8 9 10 11 12 13 14 15 16 17 18 19 20 21 22 23 24 25 26 27 28 29 30 31 32

OK, no biggie so far. This is just a list of row numbers for each of the rows of the data frame. But now look at this:

```
t.df[["Freq"]]
```

0 0 35 0 0 0 17 0 118 154 387 670 4 13 89 3 5 11 13 0 1 13 14 0 57 14 75 192 140 80 76 20

We've extracted the values in the **"Freq"** column. If we had used single brackets instead of double brackets we would have ended up with a column matrix, not a vector.

Now we need to match these two long vectors together. There were no members in the first two categories, but there are 35 in the third, so we need 35 repeats. So on until we get to the 27. We can match up the two vectors with

```
rep.int(seq_len(nrow(t.df)), t.df[["Freq"]])
```

This is exactly what we do in **as.cases**. This gives us a *list of row numbers with many repeats*. The last line of **as.cases** generates a data frame with repeating lines by extracting the rows by index.

Our case by case data frame is thus produced like this:

```
t.cases=as.cases(t.df, "Freq")
head(t.cases)
print(nrow(t.cases)}
```

Class	Sex	Age	Survived	
3	3rd	Male	Child	No
3.1	3rd	Male	Child	No
3.2	3rd	Male	Child	No
3.3	3rd	Male	Child	No
3.4	3rd	Male	Child	No
3.5	3rd	Male	Child	No

[1] 2201

Thus we have expanded our 32 line summary **t.df** into a 2201 line case by case data frame **t.cases**.

Now we return to our usual process of generating training and test sets. As usual, we will use 75% for training and keep the remaining 25% for testing.

```
n=nrow(t.cases)
ind=sample(1:n, .75*n)
test=t.cases[-ind,]
train=t.cases[ind,]
```

We can now proceed to classification.

```
model=naiveBayes(Survived~., data=train)
print{model}
```

```
Naive Bayes Classifier for Discrete Predictors

Call:
naiveBayes.default(x = X, y = Y, laplace = laplace)

A-priori probabilities:
Y
      No       Yes
0.6890909 0.3109091

Conditional probabilities:
    Class
Y            1st        2nd        3rd       Crew
  No  0.08355321 0.09850484 0.35620053 0.46174142
  Yes 0.29629630 0.16569201 0.23976608 0.29824561

    Sex
Y          Male     Female
  No  0.92084433 0.07915567
  Yes 0.49317739 0.50682261

    Age
Y         Child      Adult
  No  0.03518030 0.96481970
  Yes 0.07212476 0.92787524
```

The classifier prints the conditional probabilities for each class based upon the training data. Now we are ready to make some predictions, and we will use the teat data for this. We will evaluate our results using **confusionMatrix** from package **caret** (see page 109).

```
predictions=predict(model, test)
observed=test$Survived # column $Survived in test df
```

```
library(caret)
confusionMatrix(observed,predictions)
```

```
Confusion Matrix and Statistics

          Reference
Prediction  No Yes
       No  317  36
       Yes 113  85

              Accuracy : 0.7296
                95% CI : (0.6904, 0.7663)
   No Information Rate : 0.7804
```

```
           P-Value [Acc > NIR] : 0.9979

                         Kappa : 0.3579
         Mcnemar's Test P-Value : 4.78e-10

                   Sensitivity : 0.7372
                   Specificity : 0.7025
                Pos Pred Value : 0.8980
                Neg Pred Value : 0.4293
                    Prevalence : 0.7804
                Detection Rate : 0.5753
          Detection Prevalence : 0.6407
             Balanced Accuracy : 0.7198

              'Positive' Class : No
```

According to the confusion matrix the model has approximately a 73% accuracy, and 74% sensitivity, and a 70% Specificity. We can also generate an ROC curve (figure 14.1, but we need probabilistic predictions for that (not just yes/no). In **predict** we can get probabilities with the option **type="raw"**. There are two columns. The first column will be for **"No"** and the second for **"Yes"** (this is data dependent and depends on how the factors are defined).

```
library(pROC)
plot(roc(as.numeric(observed),
         predict(model,test,type="raw")[,2])
```

Figure 14.1.: ROC curve for Naive Bayes classification with the Titanic data.

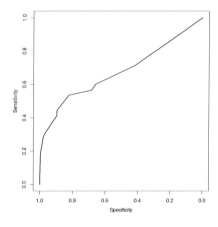

15. Discriminant Analyses

Linear and quadratic discriminant analyses are refinement on Bayesian classification.[1] To understand them we should define some standard notations. Suppose we have a response variable y that can fall into any of K classes. Then

- The **prior probabilities** $p_1, p_2, \ldots, p_k$ are some initial first guesses of the probability that an observation will fall into each of the K classes.

- The **probability density function** $f_k(x) = P(X = x|y = k)$ denotes the probability density of a variable x that is known to be drawn from class k.

Then by Bayes' theorem, the probability that an observation with explanatory variable x falls into class k is given by

$$P_k(x) = P(y = k|x) = \frac{p_k f_k(x)}{\sum_{i=1}^{K} p_i f_i(x)}$$

The goal is then to find appropriate estimates for the p_k and f_k so that we can calculate $P_k(x)$. Given $P_k(x)$ and an observation x, the class assignment is the value of k that maximizes $P_k(x)$.

Linear Discriminant Analysis

In **Linear Discriminant Analysis** (LDA), we make the following assumptions:

- The data are distribution normally;

- The explanatory variable x has the same standard deviation σ for all the classes;

---

[1] For more details the reader should refer to the texts [ISL] and [ESL]. A general review can be found in Tharwat, A. (2016) *Linear vs. quadratic discriminant analysis classifier: a tutorial.* International Journal of Applied Pattern Recognition. **3**:145-10.

- If there is more than predictor (x, feature), then the explanatory variable has the same covariance matrix Σ for all the classes.

When there is a single predictor (explanatory variable), then under these assumptions we can write

$$f_k(x) = \frac{1}{\sqrt{2\pi}\sigma} e^{-(x-\mu_k)^2/(2\sigma^2)}$$

If there are $d > 1$ features, we combine them into a vector $\mathbf{x}$ and write

$$f_k(x) = \frac{1}{\sqrt{(2\pi)^d \det \Sigma}} e^{-(1/2)(\mathbf{x}-\boldsymbol{\mu}_k)^{\mathrm{T}} \Sigma (\mathbf{x}-\boldsymbol{\mu}_k)}$$

where $\boldsymbol{\mu}_k$ is a vector of the feature means for class k. The math is simpler in one dimension, so we will continue with that. Substituting the Gaussian expression for $f_k(x)$ into the class probability P_k gives

$$P_k(x) = \frac{\dfrac{p_k}{\sqrt{2\pi}\sigma} e^{-(x-\mu_k)^2/(2\sigma^2)}}{\sum_{i=1}^{K} \dfrac{p_i}{\sqrt{2\pi}\sigma} e^{-(x-\mu_i)^2/(2\sigma^2)}}$$

We want to find the class k that maximizes P_k. Since the natural logarithm is a monotonically increasing function of its argument, $P_k(x)$ is maximized when $\ln P_k(x)$ is maximized. But

$$\ln P_k(x) = \ln p_k - \frac{(x-\mu_k)^2}{2\sigma^2} + \text{terms that do not depend on } k$$

$$= \ln p_k + \frac{\mu_k}{\sigma^2} - \frac{\mu_k^2}{2\sigma^2} + \text{terms that do not depend on } k$$

The first terms in this equation are called the **discriminant** $\delta_k(x)$,

$$\delta_k(x) = \ln p_k - \frac{(x-\mu_k)^2}{2\sigma^2}$$

When the discriminant is maximized, the probability $P_k(x)$ is maximized. The corresponding discriminant for multifactorial problems is

$$\delta_k(x) = \mathbf{x}^{\mathrm{T}} \Sigma^{-1} \boldsymbol{\mu}_k - \frac{1}{2} \boldsymbol{\mu}_k^{\mathrm{T}} \Sigma^{-1} \boldsymbol{\mu}_k + \ln p_k$$

The general procedure is to make a first guess for μ and σ. The means are calculated as the class average from the training data, and σ is a weighted average of the standard deviations. If there are n training points and K classes then (for a single factor),

$$\sigma^2 = \frac{1}{n-K} \sum_{k=1}^{K} \sum_{x_j \in C(k)} (x_j - \mu_k)^2$$

where $C(k)$ is the set of all x training values for class k. These values are substituted into the formula for δ_k, and the class with the largest discriminant is chosen. Multifactorial formulas are given in [ESL].

Figure 15.1.: LDA training set (left) and predictions (right). In the predictions plot, the training set is shown as the small markers, and the predicted points shown as the larger markers.

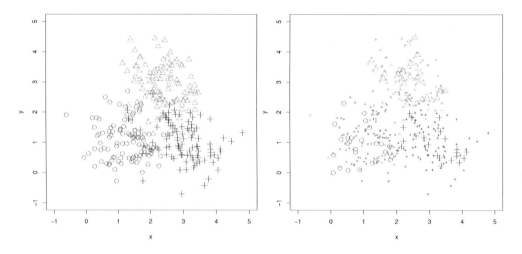

LDA can be performed with the function **lda** in the **MASS** package. We'll use **mcrnorm** (also from **MASS**) to generate some Gaussian clouds.

```
library(MASS)
set.seed(25)
M=diag(c(.4,.4))
cloud1=mvrnorm(n=100,mu=c(1.2,1.2),Sigma=M)
cloud2=mvrnorm(n=100,mu=c(2.4,2.9),Sigma=M)
cloud3=mvrnorm(n=100,mu=c(3,1),Sigma=M)
```

We can plot the three clouds with different markers (left frame of figure 15.1).

```
plot(cloud1,col="red",xlim=c(-1,5),ylim=c(-1,5),
    cex=1.5,pch=1, xlab="x",ylab="y")
points(cloud2,col="green",cex=1.5,pch=2)
points(cloud3,col="blue",cex=1.5,pch=3)
```

To do predictions we'll need to combine the clouds together into a single data frame.

```
data=rbind(cloud1,cloud2,cloud3)
colnames(data)=c("x","y")
data=data.frame(data)
```

We also need to label the three clusters. We'll create a vector with trainining labels that correspond to the rows of the data frame.

```
labels=c(rep(1,nrow(cloud1)),
        rep(2,nrow(cloud2)),
        rep(3,nrow(cloud3)))
```

The basic form of **lda** asks for a list of indices (row numbers) to use for training. We'll define these here, and then use them to extract a test set as well. The test set will be those rows that are not used for training. We'll use 75% of the data for training, and leave the rest for the test set.

```
ind=sample(1:nrow(data),.75*nrow(data))
test.data=data[-ind,]
```

Here is where we do the LDA. The first argument, **data**, is the data frame; the second argument **grouping** gives the list of training labels; and the third argument **subset** gives a list of row numbers to used for training.

```
model=lda(data,grouping=labels,subset=ind)
print{model}
```

```
Call:
lda(data, grouping = labels, subset = ind)

Prior probabilities of groups:
        1         2         3
0.3244444 0.3244444 0.3511111

Group means:
         x         y
```

Ch. 15. Discriminant Analyses

```
1 1.026256 0.8661609
2 2.498233 2.9739489
3 2.947518 1.0438932

Coefficients of linear discriminants:
        LD1         LD2
x  -1.230865  -1.1396152
y  -1.189829   0.9352518

Proportion of trace:
   LD1    LD2
0.7383 0.2617
```

Now we can make some predictions using the test data. It will e easier to make a plot if we convert the class predictions from factors to numbers.

```
predictions=predict(model,test.data)
the.classes=predictions$class # Extract $class attribute
predicted.classes=as.numeric(as.character(the.classes))
```

Here we make plot of the predicted classes, shown on the right hand frame of figure 15.1. We plot the training data as small markers (with **cex=.5**) and the predicted points as larger markers (with **cex=1.5**).

```
npred=nrow(test.data)
plot(cloud1,col="red",xlim=c(-1,5),ylim=c(-1,5),
     cex=.5,pch=1, xlab="x",ylab="y")
points(cloud2,col="green",cex=.5,pch=2)
points(cloud3,col="blue",cex=.5,pch=3)
the.colors=c("red","green","blue")
the.markers=c(1,2,3)
for (j in 1:npre){
    the.class=predicted.classes[j]
    the.point=test.data[j,]
    points(the.point, col=the.colors[the.class],
           pch=the.markers[the.class], cex=1.5)
}
```

Quadratic Discriminant Analysis

Quadratic discriminant analysis (QDA) is similar to LDA, except that instead of assuming that that the factors have have the same covariance matrix in each class, there is a different covariance matrix in each class.

This makes the equations considerably messier and we will omit them. Interested readers are referred to [ISL] and [ESL].

We'll use function **qda** from **MASS** to demonstrate QDA. Since the demonstration mimics the LDA demonstration in the previous subsection, the annotations will be very brief. You should refer back to the earlier pages for more detailed explanations of the almost identical code.

Generate (and plot) three clusters with different covariance matrices. The plot is shown in the left frame of figure 15.2.

```
set.seed(25)
cloud1=mvrnorm(n=100,mu=c(1.2,1.2),Sigma=diag(c(.7,.2)))
cloud2=mvrnorm(n=100,mu=c(2.4,2.9),Sigma=diag(c(.2,.6)))
cloud3=mvrnorm(n=100,mu=c(3.5,.5),Sigma=diag(c(.6,.6)))
plot(cloud1,col="red",xlim=c(-1,6),ylim=c(-2,5),
     cex=1.5,pch=1, xlab="x",ylab="y")
points(cloud2,col="green",cex=1.5,pch=2)
points(cloud3,col="blue",cex=1.5,pch=3)
```

Figure 15.2.: QDA training set (left) and predictions (right). In the predictions plot, the training set is shown as the small markers, and the predicted points shown as the larger markers.

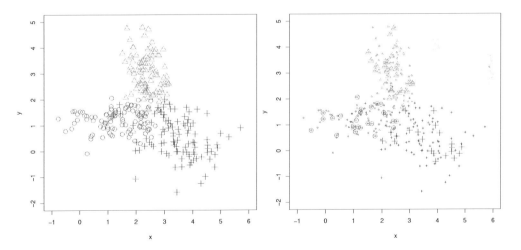

Generate training labels, and then group the clusters into a single data frame.

```
labels=c(rep(1,nrow(cloud1)),
         rep(2,nrow(cloud2)),
         rep(3,nrow(cloud3)))
data=rbind(cloud1,cloud2,cloud3)
colnames(data)=c("x","y")
data=data.frame(data)
```

Identify the training rows and test data.

```
ind=sample(1:nrow(data),.75*nrow(data))
test.data=data[-ind,]
```

Perform QDA and print the results.

```
model=qda(data,grouping=labels,subset=ind)
print(model)

Call:
qda(data, grouping = labels, subset = ind)

Prior probabilities of groups:
        1         2         3
0.3244444 0.3244444 0.3511111

Group means:
         x         y
1 1.377052  1.218566
2 2.398751  2.868094
3 3.435724  0.553758
```

Make some predictions on the test data.

```
predictions=predict(model,test.data)
the.classes=predictions$class #$class attribute
predicted.classes=as.numeric(as.character(the.classes))
```

The results can be plotted with virtually the same identical code used for the LDA. See the right hand side of figure 15.2 for the resulting plot.

16. Principal Component Analysis

PCA is not, in and of itself, a method of data classification. It is, however, sometimes for visualization and or dimension reduction. To visualize the concept of a principal component, think of a thrown American football in flight. The dynamics of spinning, oblong shaped ball are quite complicated as it twists and moves in space. Now think of making a coordinate transformation so that the x axis is fixed along the main tip-to-tip axis of the football. This is the longest axis of the football. Then cut an imaginary plane perpendicular to the x axis that is halfway between the endpoints. Place the y and z axes in this plane. Where the yz planes intersects the football we see a circle around the origin; where the xy plane intersects the football we see an oblong aligned along the x axis, centered at the origin. If we affix these axes to the football for all time, the dynamics is simple because nothing is moving. These are the principal components of the football.

The idea is to find the principal components of a large cloud of data in n dimensional space. If we can somehow order these axes so that most of the data, like the football, is sort of symmetric or spread out, along the first (or first several) axis (axes), then we may be able to ignore the projections of the data into the other dimensions. This is helpful for two reasons. First, it will allow us to more easily visualize the data – we can project into planes consisting of pairs of principal component axes. Secondly, if it is really true that only a very small percentage of the data falls into some of the dimensions, we can actually ignore those dimensions, and only study the remaining dimensions.

PCA is thus often used in combination with other techniques:

- To understand the primary directions of the spread of the data;

- To remove possibly non-useful information and simplify the data set before applying another method such as random forests;

- To visualize the data spread after clustering or other categorization techniques have been performed by projecting onto planes consisting of pairs of principal components.

Principal Components[1]

Suppose we have a collection of data points $(x_1, y_1), \ldots, (x_m, y_m)$. In higher dimensions we represent each data point by a vector **x** that we call the **feature vector**. The individual components of each feature vector are called its features. When we are dealing with xy data, the features are the x and y coordinates of each data point.

In linear regression we find a line $y = f(x)$ that minimizes the sum of the vertical distances between the points and the line. This is sometimes called the regression of y on x. This is only meaningful if all of the noise (error) is in the y-direction. If the noise is only in the x direction, we can switch coordinates and find what statisticians call the regression of x on y rather than the regression of y on x. But if there is noise along both axes what we really want is to find the line that minimizes the *perpendicular* distances to the points.

Figure 16.1.: The Principal Components of a data set. Left: Minimizing the distance to a line. Right: the Principal axes.

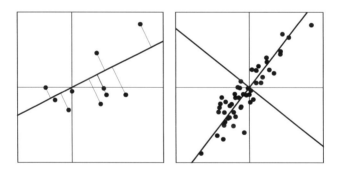

It is useful to collect together all of the data points into a single **feature matrix**. Each row of the feature matrix contains a single data point. Thus if our data consists of a collection of points in the xy plane, the feature

---

[1] The mathematical content of this section is taken from the author's text *Scientific Computation, 3rd ed.*

matrix will look like

$$\mathbf{X} = \begin{bmatrix} x_1 & y_1 \\ x_2 & y_2 \\ \vdots & \vdots \\ x_m & y_m \end{bmatrix}$$

Definition 16.1. Feature Matrix

Suppose we have a k-dimensional data set with m vectors $\mathbf{x}_1, \ldots, \mathbf{x}_m$, that we write as row vectors

$$\mathbf{x}_j = [x_{j1}, x_{j2}, \ldots, x_{jk}], \, j = 1, \ldots, m$$

Then we define the **feature matrix** $\mathbf{X}$ as a matrix whose rows are composed of the data points:

$$\mathbf{X} = \begin{bmatrix} \mathbf{x}_1 \\ \vdots \\ \mathbf{x}_m \end{bmatrix}$$

Usually we will want to have the data **zero-centered**, that is, transformed to a new coordinate system so that the average values in each axis are at the origin. If

$$\mu_x = \frac{1}{m} \sum_{j=1}^{m} x_j \text{ and } \mu_y = \frac{1}{m} \sum_{j=1}^{m} y_m$$

then the zero centered matrix (for a data set in the plane) will look like

$$\mathbf{X} = \begin{bmatrix} x_1 - \mu_x & y_1 - \mu_y \\ x_2 - \mu_x & y_2 - \mu_y \\ \vdots & \vdots \\ x_m - \mu_x & y_m - \mu_y \end{bmatrix}$$

The **scatter matrix** $\mathbf{X}^T \mathbf{X}$ occurs frequently in statistical analysis. For two dimensional data, it is computed as

$$\mathbf{X}^\mathrm{T}\mathbf{X} = \begin{bmatrix} x_1 - \mu_x & y_1 - \mu_y \\ x_2 - \mu_x & y_2 - \mu_y \\ \vdots & \vdots \\ x_m - \mu_x & y_m - \mu_y \end{bmatrix}^\mathrm{T} \begin{bmatrix} x_1 - \mu_x & y_1 - \mu_y \\ x_2 - \mu_x & y_2 - \mu_y \\ \vdots & \vdots \\ x_m - \mu_x & y_m - \mu_y \end{bmatrix}$$

$$= \begin{bmatrix} x_1 - \mu_x & \cdots & x_m - \mu_x \\ y_1 - \mu_y & \cdots & y_m - \mu_y \end{bmatrix} \begin{bmatrix} x_1 - \mu_x & y_1 - \mu_y \\ x_2 - \mu_x & y_2 - \mu_y \\ \vdots & \vdots \\ x_m - \mu_x & y_m - \mu_y \end{bmatrix}$$

$$= \begin{bmatrix} \sum(x_j - \mu_x)^2 & \sum(x_j - \mu_x)(y_j - \mu_y) \\ \sum(x_j - \mu_x)(y_j - \mu_y) & \sum(y_j - \mu_y)^2 \end{bmatrix}$$

$$= (m-1)\mathsf{cov}(\mathbf{X})$$

where $\mathsf{cov}(\mathbf{X})$ is the covariance matrix of the data set that is represented by $\mathbf{X}$. Note that this proportionality only works if the data set is centered.

Definition 16.2. Scatter Matrix

Let $\mathbf{X}$ be a centered feature matrix. Then the **scatter matrix** $\mathbf{S}$ (also called the **covariance matrix**, $\mathsf{cov}(\mathbf{X})$), is

$$\mathbf{S} = \mathbf{X}^\mathrm{T}\mathbf{X} = \mathsf{cov}(\mathbf{X})$$

Definition 16.3. Principal Direction

The **principal directions** of the centered data set are the eigenvectors of the covariance matrix $\mathbf{S}$.

Definition 16.4. Principal Components

The **principal components** of a data set $\mathbf{X}$ are the projections of the data set onto its principal directions.

In particular, once you know the principal directions $\mathbf{v}_i$ of a centered data set $\mathbf{X}$, then the principal components of that data set are the projections $\mathbf{Xv}_i$. If the $\mathbf{v}_i$ are the column vectors of a matrix $\mathbf{V}$, then we compute $\mathbf{XV}$. This is sometimes used for data set **dimensionality reduction**. It is not unusual for a data set to be oriented in such a way that many of its principal components are dominant. We can choose the projection of the data set into the subspace represented by the dominant components to get a lower dimensional representation that contains nearly all of the information that is contained in the original data set.

> **Theorem 16.1. Principal Component Calculation Theorem**
>
> Let $\mathbf{X}$ represent a zero centered data set with singular value decomposition $\mathbf{X} = \mathbf{U}\Sigma\mathbf{V}^T$. Then the principal directions of $\mathbf{X}$ are the column vectors of $\mathbf{V}$.

We expand the matrix $\mathbf{S} = \mathbf{X}^T\mathbf{X}$ using its singular value decomposition,

$$\mathbf{S} = \mathbf{X}^T\mathbf{X} = (\mathbf{U}\Sigma\mathbf{V}^T)^T(\mathbf{U}\Sigma\mathbf{V}^T)$$

Since the transpose of a product is the product of the transposes in the reverse order,

$$\mathbf{S} = \mathbf{V}\Sigma\mathbf{U}^T\mathbf{U}\Sigma\mathbf{V}^T$$

Since $\mathbf{U}$ is is orthogonal, then $\mathbf{U}^T\mathbf{U} = \mathbf{I}$ (the identity matrix), and thus

$$\mathbf{S} = \mathbf{V}\Sigma^2\mathbf{V}^T$$

Right multiplying by $\mathbf{V}$,

$$\mathbf{SV} = \mathbf{V}\Sigma^2\mathbf{V}^T\mathbf{V} = \mathbf{V}\Sigma^2\mathbf{I} = \mathbf{V}\Sigma^2 end$$

because $\mathbf{V}$ is orthogonal. Since Σ is diagonal, so is Σ^2. If s_i, $i = 1, 2, \ldots, k$ are the numbers on the diagonal of Σ, then $s_1^2, s_2^2, \ldots, s_k^2$ are on the diagonal of Σ^2.

Let $\mathbf{v}_1, \mathbf{v}_2, \ldots, \mathbf{v}_k$ be the column vectors of $\mathbf{V}$. Then $\mathbf{SV} = \mathbf{V}\Sigma^2$ becomes

$$\mathbf{S}\begin{bmatrix} \mathbf{v}_1 & | & \mathbf{v}_2 & | & \cdots & | & \mathbf{v}_k \end{bmatrix} = \begin{bmatrix} \mathbf{v}_1 & | & \mathbf{v}_2 & | & \cdots & | & \mathbf{v}_k \end{bmatrix} \begin{bmatrix} s_1^2 & 0 & \cdots & 0 \\ 0 & s_2^2 & & \vdots \\ \vdots & 0 & \ddots & 0 \\ 0 & \cdots & 0 & s_k^2 \end{bmatrix}$$

$$= \begin{bmatrix} s_1^2\mathbf{v}_1 & | & s_2^2\mathbf{v}_2 & | & \cdots & | & s_k^2\mathbf{v}_k \end{bmatrix}$$

Equating the columns,

$$\mathbf{S}\mathbf{v}_j = s_j^2\mathbf{v}_j, \; j = 1, 2, \ldots, k$$

Thus the s_j^2 are the eigenvalues of $\mathbf{S}$ and the $\mathbf{v}_j$ are the corresponding eigenvectors. It is customary to order these so that $s_1^2 \geqslant s_2^2 \geqslant \cdots \geqslant s_k^2$.

Algorithm 16.1 Principal Components of a Data Set.

---
input: Data set represented by zero-centered matrix $\mathbf{X}$
1: $\mathbf{U}, \Sigma, \mathbf{V}^\mathrm{T} \leftarrow \mathrm{SVD}(\mathbf{X})$
2: **return** $\begin{bmatrix} \mathbf{v}_1 & | & \mathbf{v}_2 & | & \cdots & | & \mathbf{v}_k \end{bmatrix}$ (column vectors of $\mathbf{V}$)

PCA in R

We will examine a table of cost data that compares the cost of gasoline (in dollars per gallon) and electricity (in dollars per kilowatt hour) that was extracted from the US Bureau of Labor Statistics web site.[2] The data covers the period from 2004 through 2014.

```
gecost=read.csv("gasgal-vs-eleckwh.csv",
    header = TRUE, sep = ",")
print(head(gecost, 5))
```

```
  gas.per.gallon elec.per.kwh
1          1.592        0.091
2          1.672        0.091
3          1.766        0.091
4          1.833        0.091
5          2.009        0.093
```

For comparison, we will fit a linear model to the data.

```
lmodel=lm(elec.per.kwh~gas.per.gallon, gecost)
lmodel
```

---
[2] The web site is at https://data.bls.gov and the data was downloaded in 2016. The data does not seem to be posted there any longer, or has been moved to another location that is not easily identified.

```
Call:
lm(formula = elec.per.kwh ~ gas.per.gallon, data = gecost)

Coefficients:
   (Intercept)   gas.per.gallon
       0.07605          0.01541
```

The least squares linear regression to this data is

$$y = 0.07605 + 0.01541x$$

We can compute the principal components with the function **prcomp**. This is not the only function in R that can compute principal components, but it is built into the **stats** package which is always loaded automatically, so you don't have to install or load anything.

```
pc=prcomp(gecost)
print(pc)
```

```
Standard deviations (1, .., p=2):
[1] 0.65108319 0.00927744

Rotation (n x k) = (2 x 2):
                      PC1          PC2
gas.per.gallon  0.99988131  -0.01540671
elec.per.kwh    0.01540671   0.99988131
```

The principal components are vectors pointing along lines in the direction of the main spread of the data. If we place them at the center of the data, they will be along axes of symmetry of an ellipsoid that the represents the data. We can extract the vectors from the **rotation** part of the PCA and calculate the slop of each of them.

```
PC1=pc$rotation[,1]
PC2=pc$rotation[,2]
slope1 = as.numeric(PC1[2]/PC1[1])
slope2 = as.numeric(PC2[2]/PC2[1])
c(slope1,slope2)
```

0.0154085366224698 -64.8990896735598

Consider the first component, with slope $b \approx 0.01541$. A line $y = a + bx$ that passes through the point (x_1, y_1) will satisfy the equation $y_1 = a + b_x$ or $a = y_1 - bx_1$. Lets use the column means for a point so that we find this line. The column means are given by

```
means=as.numeric(colMeans(gecost))
yc=means[2]; xc=means[1]
means
```

2.91208333333333 0.120916666666667

Thus the intercept a is given by

```
intercept=yc - slope1 * xc
intercept
```

0.076045723977316

The line passing the center and in the direction of the first principal component is (rounded to 5 decimals),

$$y = 0.07605 + 0.01541x$$

This is the same as the least squares linear fit.

With the definitions of **xc**, **yc**, **PC1**, and **PC2** above, we can draw short arrows on a scatter plot of the data to illustrate the principal components. We have to tune the length of the arrows manually to make the plot "pretty." We'll also use **abline** to plot the line along the first principal component (figure 16.2).

```
library{graphics} # for arrows
plot(gecost$gas.per.gallon, gecost$elec.per.kwh, cex=0.5,
    main="Gasoline and Electricity Cost, 2004-2014",
    xlab="Gasoline, Dollars/gallon",
    ylab="Electricity, Dollars/kwh", ylim=c(0.07,.175))
arrows(xc, yc, xc+PC1[1]*.3, yc+PC1[2]*.3, lwd=5)
arrows(xc, yc, xc+PC2[1]*.025, yc+(PC2[2]*.025), lwd=5)
abline(intercept, slope1)
```

Image Compression

By treating an image as a matrix of pixels allows us to find its principal components through the SVD. This produces one possible image compression scheme (although it is not without some information loss). We'll start by reading in a **jpg** image.

Figure 16.2.: Scatterplot of the electricity versus gasoline cost data, showing the two principal components.

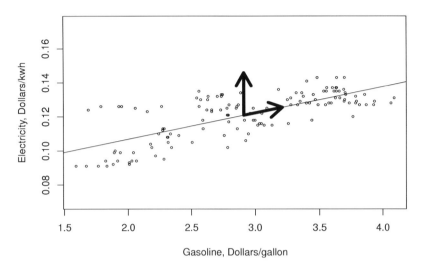

```
# install.packages("jpeg")
library(jpeg)
romeo=readJPEG("my-bone.jpg")
print(dim(romeo))
```

2080 1536 3

The jpg file consists of three matrices of pixels. Each one is $2080x1536$ and represents the red, blue and green layers in the RGB image. Each pixel is represented by a real number in the range $[0, 1]$. We will convert this jpg into a gray scale image using the averaging method.

```
red=romeo[,,1]
green=romeo[,,2]
blue=romeo[,,3]

gray=(red+blue+green)/3
```

Each pixel in **gray** is still between 0 and 1. The procedure we are demonstrating for the gray scale image is also valid for RGB images, but each layer must be processed separately and the results combined. For simplicity, it is easier to describe the process with only a single layer.

Because of the way image files are implemented in R, when they get converted to a matrix, the data gets rotated by 90 degrees. This does not affect the results, but it makes it more difficult to visualize. For presentation purposes, it is more convenient to look at the image in its original format. Consider the following array:

$$\begin{array}{cccc} a & b & c & d \\ e & f & g & h \\ i & j & k & l \end{array}$$

Suppose we reverse the order of all the columns. Then we have:

$$\begin{array}{cccc} i & j & k & l \\ e & f & g & h \\ a & b & c & d \end{array}$$

If we now take the transpose, we end up with the original matrix rotated by 90 degrees.

$$\begin{array}{ccc} i & e & a \\ j & f & b \\ k & g & c \\ l & h & d \end{array}$$

We can reverse the order of a vector with **rev** function. To apply this to all the columns in a matrix **X** we use the function **apply(X, 2, rev)**. The second argument (in this case **2**), tells **apply** to the columns. Had we set the second argument to **1**, the rows would have been reversed instead of the columns. To reverse the columns and then take the transpose we can use the transpose function **t(apply(X, 2, rev))**.

```
gray = t(apply(gray, 2, rev))
```

Once an image is represented by a matrix, we can display it on the screen with the **image** function. The gray scale image (bottom right of figure 16.3) can be displayed, for example, with:

```
image(gray, col=gray.colors(256), axes=FALSE)
```

Compute the SVD (see appendix A), and separate the **S**, **U**, and **V** matrices, where the SVD is $\mathbf{USV}^T$.

```
gray.svd=svd(gray)

S=gray.svd$d    #$d = sing. values
U=gray.svd$u    #$u = U matrix
V=gray.svd$v    #$v = V matrix
```

Here the columns of **svd$v** are the principal components of the principal component decomposition. Using theorem A.13,

$$\mathbf{X} = \mathbf{USV}^T = S_1 \mathbf{u}_1 \mathbf{v}_1^T + S_2 \mathbf{u}_2 \mathbf{v}_2^T + \cdots + S_n \mathbf{u}_n \mathbf{v}_n^T$$

If there are **K** principal components, we can thus expand the image representation as

```
I = 0
for (j in 1:K){
   I = I +  S[j] * ((U[,j]) %o% t(V)[j,])}
```

If we truncate the sum at a lower value we are only using some of the principal components. The idea is to use the most important components. Since the SVD software (as well as the PCA software) arranges the components in order, by decreasing value of the singular values. Thus the components that are the most significant are included first. These components have the most information. The components later in the list have less information.

We can write a function **ShowImageComponents** that calculates the sum of the first $n \leq K$ components. We need to perform the SVD prior to calling **ShowImageComponents**.

```
ShowImageComponents = function(n, S, U, V){
   total=0
   for (j in 1:n){
      total = total +  S[j] * ((U[,j]) %o% t(V)[j,])
   }
   w=dim(V)[1]; v=dim(V)[2]
   pic=matrix(total)
   dim(pic)=c(1,w)
   pic
}
```

We can write an image-specific function so that we can compare the results of different truncations. Our new function **ShowRomeo** invokes the previously defined function **ShowImageComponents** using the values of **S**, **U**, **V** that we have calculated from our picture of Romeo.

```
ShowRomeo = function(n){
    I= ShowImageComponents(n, S, U, V)
    image(I, col=gray.colors(256),
        #xlab=paste(n, "components"),
        axes=FALSE)
}
```

We can compare various truncations (see figure 16.3) with

```
ShowRomeo(10)          # First 10 principal components
ShowRomeo(25)          # First 25 principal components
ShowRomeo(50)          # First 50 principal components
ShowRomeo(100)         # First 100 principal components
ShowRomeo(1000)        # First 1000 principal components
```

This can result in substantial space savings if not all of the components are kept, because only the eigenvectors and eigenvalues need to stored, and not the entire matrix. For this image, for example, the corresponding vectors have length 1536. As there are 2080 of them, storing the entire image requires $1536 \times 2080 = 3,194,880$ words of storage. If the truncation is terminated after n components, it only requires $2n \times 1536 + n$ words. (The 2 is because both U and V vectors, and the second n is for the eigenvalue.) For example, if we only keep 100 components, then $2 \times 100 \times 1536 + 100 = 307{,}300$ words are required.

Figure 16.3.: Compressed images. Top row, left to right: 10, 25, and 50 principal components. Bottom row, from left: 100, and 1000 components. Bottom, far right: original image after gray scale conversion.

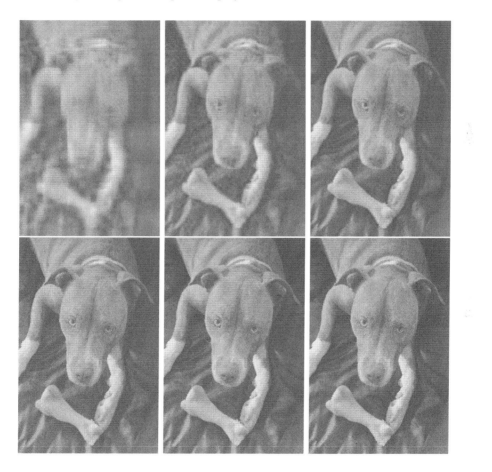

17. Support Vector Machines

In the general clustering problem we have a collection of N points $\mathbf{x}_i$, $i = 1, \ldots, N$ in d-dimensional space, and we want to group them into K different clusters. In a *linear separation* problem, we would find planes that separate the clusters. An example is given by k-means clustering, where a plane is found that is halfway between each pair of cluster centers. Consider the two dimensional case, where we usually write each point in a plane as a pair (x, y). Since we are not treating y as anything special, we can write the coordinates of a single point sa $\mathbf{x} = (x_1, x_2)$, with the obvious generalization to d dimensional space of $\mathbf{x} = (x_1, \ldots, x_d)$. A line

$$y = c_0 + bx$$

then becomes

$$x_2 = c_0 + c_1 x_1,$$

where we have written the slope $b = c_1$ and the y-intercept a as c_0. The d-dimensional generalization is

$$x_d = c_0 + c_1 x_1 + c_2 x_2 + \cdots + c_{d-1} x_{d-1}.$$

Upon rearrangement we have for each equation

$$0 = c_0 + c_1 x_1 + c_2 x_2$$

in the plane, where $c_2 = -1$ is a fixed constant; and

$$0 = c_0 + c_1 x_1 + c_2 x_2 + \cdots + c_d x_d$$

in d dimensions, where $c_d = -1$ is a fixed constant. The d-dimensional generalization of the line is called a **hyperplane in d dimensions**. If we defined the **weight vector**

$$\mathbf{c} = (c_1, c_2, \ldots, c_d)^\mathrm{T}$$

where $d = 2$ for the usual 2-dimensional line, and $d > 2$ for a hyperplane in more than two dimensions, then we have, **in either case**,

$$0 = c_0 + \mathbf{c}^\mathrm{T} \mathbf{x}$$

Figure 17.1.: Support vectors for two different lines that linearly separate the same pair of clusters.

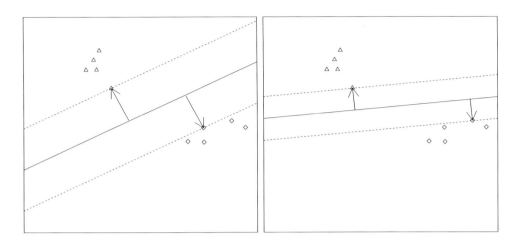

If the line $y = a + bx$, which we have rewritten as $c_0 + \mathbf{c}^T\mathbf{x} = 0$ separates the two clusters, then the points in either cluster can be identified because

$$c_0 + \mathbf{c}^T\mathbf{x} \begin{cases} > 0 & \text{points in cluster 1} \\ = 0 & \text{points on the line} \\ < 0 & \text{points in cluster 2} \end{cases}$$

The vector $\mathbf{c}$ is perpendicular to the line $c_0 + \mathbf{c}^T\mathbf{x} = 0$ (figure 17.1). To see this, consider the two dimensional case (line $y = a + bx$). The we can write

$$\mathbf{c} = (c_1, c_2) = (b, -1)$$

Similarly, a vector v parallel $y = a + bx$ would have a change in y equal to the slope b for every change in x of 1, i.e.,

$$\mathbf{v} = (1, b)$$

Since

$$\mathbf{v} \cdot \mathbf{c} = 0$$

the vectors are perpendicular.

Suppose the line $y = a + bx$ is drawn so that it separates two clusters, i.e., one cluster is completely on one side of the line and the other cluster is on the other side. Next, consider sweeping the vector $\epsilon\mathbf{c}$ along the line $y = a + bx$. At some place along the line, $\epsilon\mathbf{c}$ will point directly at a point. The smallest distance

$$M = \min_j |\epsilon\mathbf{c}|$$

where the minimum is taken over all the points in either cluster is called the **support**.

Different lines that separate the clusters will have different supports. The idea of a **support vector machine** is to find a line with the largest support that makes the fewest number of mistakes in classification. The equations can be solved using the calculus of variations (see, for example, [ESL]) and are not given here.

Figure 17.2.: Results of **svm** for separating two overlapping Gaussian clusters in R. On the left, the entire data set is shown. On the right, training data is shown with small markers, predictions for test data with large markers.

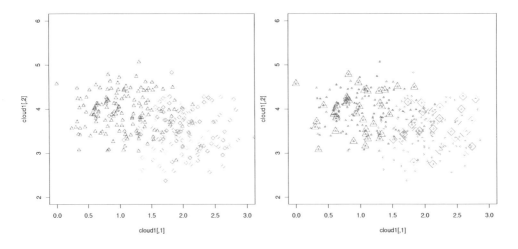

To demonstrate how to find the support vectors in R, we will generate some toy data in two dimensions. We will use the Gaussian random number generator **mvrnorm** in **MASS** to generate to clouds of overlapping data. The data is plotted in the left frame of figure 17.2.

```
library(MASS)
set.seed(999)
ncloud=150
cloud1=mvrnorm(n=ncloud,mu=c(1,4),Sigma=diag(c(.2, .2)))
cloud2=mvrnorm(n=ncloud,mu=c(2,3.5),Sigma=diag(c(.2, .2)))
plot(cloud1,xlim=c(0,3),ylim=c(2,6),pch=2,col="red")
points(cloud2,pch=5,col="green")
xy=rbind(cloud1,cloud2)
colnames(xy)=c("x","y")
labels=as.factor(c(rep(1,ncloud),rep(2,ncloud)))
```

The function we need to use to calculate the support vectors is in the package **e1071** (see the footnote on page 131).

```
library(e1071)
```

Here we will separate the data into test and training data. In addition we will extract the the training labels.

```
nclouds=2*ncloud
ind=sample(1:nclouds,.75*nclouds)
train=xy[ind,]
test=xy[-ind,]
train.labels=labels[ind]
```

The function for computing the the support vectors is **svm**. It requires two inputs: a data frame containing the test data; and a list of labels identifying the clusters.

```
model=svm(train,train.labels)
print(model)
```

```
Call:
svm.default(x = train, y = train.labels)


Parameters:
   SVM-Type:  C-classification
 SVM-Kernel:  radial
       cost:  1
      gamma:  0.5

Number of Support Vectors:  82
```

Ch. 17. Support Vector Machines

The function **predict** will take as input a model produced by **svm** and a new data frame and produce its predictions as a factor matrix. To plot the factors we need to convert them to numbers.

```
predictions=as.numeric(as.character(predict(model,test)))
```

Finally, we'll plot both the training data and the predicted cluster for the test data on a single figure. Well use small markers for the training data and larger markers for the test data. The result is shown on the right hand side of figure 17.2.

```
plot(cloud1,xlim=c(0,3),ylim=c(2,6),pch=2,col="red",cex=.5)
points(cloud2,pch=5,col="green",cex=.5)
cloud.colors=c("red","green")
cloud.markers=c(2,5)
ntest=nrow(test)
for (j in 1:ntest){
    pj=predictions[j]
    points(test[j,1], test[j,2],
            col=cloud.colors[pj],
            pch=cloud.markers[pj],
            cex=2)
}
```

18. Decision Trees

As with regression trees (see chapter 9), decision trees are formed by recursively partitioning the domain of the predictor variable.[1] In most implementations the tree is binary, i.e., a hierarchy of partitions is formed in which regions are recursively divided into two sub-nodes. The resulting tree is rarely balanced, hence the sub-tree from the right child may be much deeper than than the sub-tree from the left child. As the regions associated with successive partitions become smaller, they will contain fewer predictors. When an appropriate stopping condition is reached, that particular branch will cease to split and the predictors in the final node (called a leaf) are assigned to a class based on the target classes (the y values assigned to the x values in the data set that fall into the corresponding region). This differs from a regression tree, in which the predicted y values are assigned a valued based on the average of the target values (the actual y-values).

The decision on where to partition a region is called a **splitting criterion**. For multivariate data, each variable must be considered at each split, and a decision is made to determine the optimal split. The optimal split tells us (a) which variable to use; and (b) where to place the cut point. Many different methods can be used to make this decision. There is no best method; like everything else in ML, it generally depends on the data set. Here are some of the more common methods:

- Information Gain. The entropy is

$$S = -\sum_i p_i \log p_i$$

  where p_i is the proportion assigned to each class for a given split, the logarithm is calculated in base 2, and we define $p \log p = 0$ when $p = 0$. The corresponding information gain is obtained from the individual entropies.

$$\text{Information Gain} = -\sum_i \frac{|S_i|}{|S|} S_i$$

---

[1] For a more detailed survey of the mathematics: Rokach, L, and Maimon, O. *Top-Down Induction of Decision Trees. Classifiers – A Survey*. IEEE Trans. on Systems, Man, and Cybernetics – Part C. **35**(4): 476-486.

where the sum is over all the features.

- Gini Impurity. For each feature k (variable) let n_i be the number of data points that belong to class i. If there are n total data points then if we define the proportion $p_i = n_i/n$,

$$G = 1 - \sum_{i=1}^{K} p_i^2$$

  where K is the number of classes.

- Total RSS - see chapter 9.

- Chi squared, where

$$\chi^2 = \sum_i \frac{(y_i - \hat{y}_i)^2}{\hat{y}_i}$$

  A higher chi-squared value gives more information.

Different implementations use different stopping criteria, and most use some combination. The most common ones set a minimum number of points in each leaf node and a maximum tree depth.

There is a trade-off between stopping condition and tree size. If the stopping condition is two strict, the tree will be very small and not very useful. If the stopping condition is not very weak, there is a high likelihood of overfitting the data. One way to address this is to use pruning methods. The idea is to overfit the data and then selectively cut back parts of the tree in a way that the fit will become "more optimal.' A variety of algorithms employ a number of pruning methods to cut back trees in this way.

We can demonstrate decision trees using **rpart** using the wine quality data set in the UCI machine learning data archive. There are two files in the archive, one for red wine and one for white wine. There are 1550 records in the red wine file. Each wine is rated on a scale of 1 through 9. In addition, for each wine, eleven other physically quantifiable items are tabulated in the file (fixed acidity, volatile acidity, citric acid, residual sufar, chlorides, free sulfur dioxide , total sulfur dioxide, density, pH, sulphates, alcohol percentage). We load the file directly from the web site into a local variable that we will call **red.wine.data**.

```
wineurl="http://archive.ics.uci.edu/ml/machine-
learning-databases/wine-quality/winequality-red.csv"
red.wine.data=read.csv(wineurl, header = TRUE, sep = ";")
```

Note that even though the url is spread to two lines to fit on the pages of this book, it must be typed on a single line in R.

As usual, we partition the data into test and training sets. We will reserve 25% for testing.

```
n=nrow(red.wine.data)
training.indices=sort(sample(1:n,.75*n))
train=red.wine.data[training.indices,]
test=red.wine.data[-training.indices,]
```

We will build our decision tree using the function **rpart** using the training set. We must specify which features (x-variables) it depends on, and we do this by giving **rpart** a formula. Only linear terms are allowed. We tell it to fit **quality** as a function of all other columns. Option **method="class"** tells **rpart** that we are building a decision tree and not a regression tree.

```
library(rpart)
wine.model=rpart(quality~., train,  method="class"))
```

We can use **rpart.plot** to make a plot.

```
library{rpart.plot}
rpart.plot{wine.model}
```

The decision tree produced by **rpart.plot** is shown in figure 18.1.

We will use the model to make predictions with the test data using the function **predict**.

```
predictions=predict(wine.model, test, type="class")
```

To see how well our model worked on the test data we will generate the multi-class confusion matrix. The function **confusionMatrix** requires the following libraries (which must be installed if you you don't already have them.

Figure 18.1.: Plot of decision tree on wine quality using default parameters.

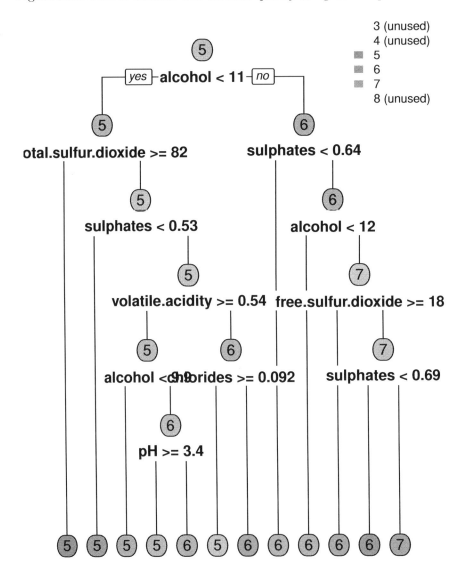

```
install.packages("e1071")
install.packages("carettree",
   repos='http://cran.us.r-project.org')
library(caret)
```

The call to **RCODE** requires that both the observations and the predictions be converted to a data structure called a **factor**. There is a function **as.factor** that will do this for us.

```
confusionMatrix(
   as.factor(predictions),
   as.factor(test$quality)) # $quality column of test
```

```
Confusion Matrix and Statistics

          Reference
Prediction   3   4   5   6   7   8
         3   0   0   0   0   0   0
         4   0   0   0   0   0   0
         5   1   7 103  39   2   0
         6   0   4  61 125  40   6
         7   0   0   0   3   6   3
         8   0   0   0   0   0   0

Overall Statistics

               Accuracy : 0.585
                 95% CI : (0.535, 0.6337)
    No Information Rate : 0.4175
    P-Value [Acc > NIR] : 1.179e-11

                  Kappa : 0.3017
 Mcnemar's Test P-Value : NA

Statistics by Class:

                     Class: 3 Class: 4 Class: 5 Class: 6 Class: 7 Class: 8
Sensitivity            0.0000   0.0000   0.6280   0.7485   0.1250   0.0000
Specificity            1.0000   1.0000   0.7924   0.5236   0.9830   1.0000
Pos Pred Value            NaN      NaN   0.6776   0.5297   0.5000      NaN
Neg Pred Value         0.9975   0.9725   0.7540   0.7439   0.8918   0.9775
Prevalence             0.0025   0.0275   0.4100   0.4175   0.1200   0.0225
Detection Rate         0.0000   0.0000   0.2575   0.3125   0.0150   0.0000
Detection Prevalence   0.0000   0.0000   0.3800   0.5900   0.0300   0.0000
Balanced Accuracy      0.5000   0.5000   0.7102   0.6361   0.5540   0.5000
```

Another way to visualize the result is to count the observed and predicted number of wines in each category. The **table** function does this for us. For the total data set,

```
TE=table(red.wine.data$quality)  # $ column quality
TE
```

```
  3   4   5   6   7   8
 10  53 681 638 199  18
```

For the test set,

```
TP=table(predictions)
TP
```

```
  3   4   5   6   7   8
  0   0 152 236  12   0
```

Since the total numbers in each group are different, we are really comparing apples and oranges. We really need to calculate the *proportion* of wines in each class.

```
P=sum(TP)
TP/P
```

```
   3    4    5    6    7    8
0.00 0.00 0.38 0.59 0.03 0.00
```

```
E=sum(TE)
round(TE/E,2)
```

```
   3    4    5    6    7    8
0.01 0.03 0.43 0.40 0.12 0.01
```

As we can see, this particular tree is not very accurate – there are a lot of off-diagonal entries in the confusion matrix. One way to try to improve this is to vary the control parameters of **rpart**. For example, you can try to decrease the complexity parameter until the tree is very deep and then prune back some branches. Other ways that can produce substantial improvement are bagging (chapter 19), boosting and random forests as ways of improving decision tree algorithms.

19. Bagging

In **ensemble learning**, multiple different models are developed and then some sort of averaging technique is used to combine them together. The simplest way of doing this is to apply the *same* algorithm to a several different training sets and then take the average prediction. Sometimes we don't have enough data to create to multiple training sets, so instead we take multiple samples from the same training set and pretend that each sample is a different training set. Since the sampling is done with replacement, it is likely that the simulated training sets will have some overlap. This is the essence of **bagging**.[1] Some authors refer to bagging as **bootstrap aggregation**.

Bagging for Decision Trees

We repeat the example presented in chapter 18, but will do bagging instead. We will use the package **ipred** to do the bagging.[2] This package implements bagging using the **rpred** function for decision trees. The main function is **ipredbagg**.

We begin by loading the data set. We could type the entire URL into a single string if we wanted, but to fit on the book page we divide it up into three lines and paste them together

```
uci="http://archive.ics.uci.edu/"
mld="ml/machine-learning-databases/"
redcsv="wine-quality/winequality-red.csv"
wineurl=paste(uci,mld,redcsv,sep="")
#
red.wine.data=read.csv(wineurl, header = TRUE, sep = ";")
```

The next step is to build test and training sets. We will pick 75% of the

---

[1] Breiman, L (1996) *Bagging predictors*. Machine Learning. **24**:123-140.
[2] Detailed descriptions of the functions in this package are given in the reference manual https://cran.r-project.org/web/packages/ipred/ipred.pdf. Additional examples can be found in the vignette *Some more or less useful examples for illustration*, at https://cran.r-project.org/web/packages/ipred/vignettes/ipred-examples.pdf.

data for training and 25% for test.

```
n=nrow(red.wine.data)
training.indices=sort(sample(1:n,.75*n))
train=red.wine.data[training.indices,]
test=red.wine.data[-training.indices,]
ntrain = nrow(train)
```

To create a decision tree the y data needs to be converted to factors. The x data is taken from the first 11 columns. The y data is in the 12th column and corresponds to the column header **"quality"**. We can extract the column by keyword (using the dollar sign notation) or column index.

```
y.train=as.factor(train[,12])
x.train=train[,1:11]

y.test=as.factor(test[,12])
x.test=test[,1:11]
```

Load the library **ipred** and make a model using the *training* data.

```
library(ipred}
bag.model = ipredbagg(
   y.train, x.train,
   nbagg=25,
   control=
      rpart.control(minsplit=2, cp=0, xval=0),
   comb=NULL, coob=FALSE, ns=ntrain, keepX = TRUE)
bag.model
```

`Bagging classification trees with 25 bootstrap replications`

We can make predictions using the *test* data set and the model. We convert the output, which is a list of values in quality space, to a factor format. We force the factors to have the same levels as the test set, just in case some of the factors are missing. This will mean that we can easily compare the expected and predicted data in the next step.

```
bag.predict= predict(bag.model, test)
bag.predict=factor(bag.predict, levels=levels(y.test))
```

We can visualize the results for the population by taking histograms of the observed and predicted tables.

```
options(repr.plot.width=7, repr.plot.height=4)
barplot(c(table(bag.predict)/sum(table(bag.predict)),
        table(y.test)/sum(table(y.test))),
        space=0,col=c(rep("red", 6), rep("white",6)))
legend("center", legend=c("expected", "predicted"),
       fill=c("red","white"))
```

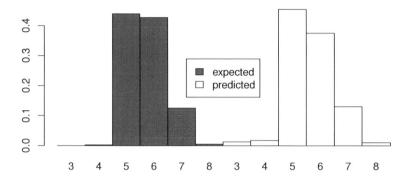

To print the confusion matrix,

```
library(caret)
confusionMatrix(data=bag.predict, reference=y.test)
```

```
Confusion Matrix and Statistics

          Reference
Prediction   3   4   5   6   7   8
         3   0   0   0   0   0   0
         4   0   0   1   0   0   0
         5   4   4 134  31   3   0
         6   1   3  43 104  17   3
         7   0   0   4  14  31   1
         8   0   0   0   1   1   0

Overall Statistics

               Accuracy : 0.6725
                 95% CI : (0.6241, 0.7183)
    No Information Rate : 0.455
    P-Value [Acc > NIR] : < 2.2e-16

                  Kappa : 0.4744
 Mcnemar's Test P-Value : NA

Statistics by Class:

                     Class: 3 Class: 4 Class: 5 Class: 6 Class: 7 Class: 8
Sensitivity            0.0000   0.0000   0.7363   0.6933   0.5962   0.0000
Specificity            1.0000   0.9975   0.8073   0.7320   0.9454   0.9949
Pos Pred Value            NaN   0.0000   0.7614   0.6082   0.6200   0.0000
Neg Pred Value         0.9875   0.9825   0.7857   0.7991   0.9400   0.9899
Prevalence             0.0125   0.0175   0.4550   0.3750   0.1300   0.0100
Detection Rate         0.0000   0.0000   0.3350   0.2600   0.0775   0.0000
Detection Prevalence   0.0000   0.0025   0.4400   0.4275   0.1250   0.0050
Balanced Accuracy      0.5000   0.4987   0.7718   0.7127   0.7708   0.4975
```

There is some improvement over a single decision tree.

20. Boosting

In boosting, a classification algorithm is iteratively improvement by classifying on the residual errors. In the most popular implementation, adaboost (adaptive boosting), classification trees are built and combined in a weighted algorithm where the weights decrease with time (iteration) so that successive iterations have less impact on the result.[1]

Boosting Decision Trees

We can illustrate boosting using the **adabag** package. This includes packages for both bagging and boosting using the adaboost algorithm.[2]

We will repeat the example used in chapter 18 by again examining the red wine quality file from the UCI machine learning archive. The following code illustrates how to combine three short strings into a long string using the **paste** command. In fact, you can type the entire url manually into a single string in your code if you like, but you may not include any line feeds (line wraps) in the url.

```
uci="http://archive.ics.uci.edu/"
mld="ml/machine-learning-databases/"
redcsv="wine-quality/winequality-red.csv"
wineurl=paste(uci,mld,redcsv,sep="")
wineurl
```

'http://archive.ics.uci.edu/ml/machine-learning-databases/wine-quality/winequality-red.csv'

The last line was included only for human verification of pasting success.

We can load the file and then print out the column headers to remind ourselves of the variable names.

---

[1] Y. Freund, R. Schapire (1999) *A Short Introduction to Boosting.* J. Jap. Soc. Art. Int. 14(5):771-780.

[2] Detailed information of **adabag** can be found at https://cran.r-project.org/web/packages/adabag/index.html in the reference manual https://cran.r-project.org/web/packages/adabag/adabag.pdf.

```
red.wine.data=read.csv(wineurl, header = TRUE, sep = ";")
colnames(red.wine.data)
```

```
'fixed.acidity' 'volatile.acidity' 'citric.acid'
'residual.sugar' 'chlorides' 'free.sulfur.dioxide'
'total.sulfur.dioxide' 'density' 'pH' 'sulphates'
'alcohol' 'quality'
```

The last column, **'quality'**, will be the response variable. All of the other columns will be predictor variables.

Create training and test sets. We will use 75% of the data for training and the rest for test.

```
n=nrow(red.wine.data)
training.indices=sort(sample(1:n,.75*n))
train=red.wine.data[training.indices,]
test=red.wine.data[-training.indices,]
ntrain = nrow(train)
```

Finally, we replace the **quality** column in the training set with a column of **factor** types. We have to do this because **factor**'s are required by the **boosting** implementation.

```
train$quality = as.factor(train$quality)
```

Load the necessary packages. Both **rpart** and **adabag** are needed for the boosting implementation, and **carat** is needed for the confusion matrix.

If this is the first time we have used the **adabag** package, we have to install the package as well. It is commented out to remind you that the package only needs to be installed once and never again.

```
#install.packages("adabag",
#   repos='http://cran.us.r-project.org')
#
library(rpart)
library(adabag)
library(caret)
```

We can do the boosting using default parameter values suggested by the documentation. Modifying these parameters may change the quality of your convergence.

```
adaboost.model = boosting(
    quality~.,      # formula, same format as lm
    data=train,     # data frame
    boos=TRUE,
    coeflearn="Breiman",
    mfinal = 100,   # number of iterations
    par=FALSE,      # not parallel processor
    control=rpart.control(maxdepth=10, minsplit=15))
```

To test the fit, we make some predictions using the test data and then calculate the confusion matrix.

```
boost.predict=predict(adaboost.model, newdata=test)
summary(boost.predict)
```

```
          Length Class   Mode
formula   3      formula call
votes     2400   -none-  numeric
prob      2400   -none-  numeric
class     400    -none-  character
confusion 18     table   numeric
error     1      -none-  numeric
```

We need to convert the predictions to a **factor** before it can be used by **confusionMatrix**. The actual prediction values are in the column **$class** of **boost.predict**. We'll print the first 10 values just to see what they look like.

```
pred=factor(boost.predict$class, # get $class
    level=levels(y.test))
print(pred[1:10])
```

```
 [1] 5 5 5 5 6 5 5 5 5 6
Levels: 3 4 5 6 7 8
```

```
confusionMatrix(data=pred, reference=y.test)
```

```
          Reference
Prediction   3   4   5   6   7   8
         3   0   0   0   0   0   0
         4   0   0   0   0   0   0
         5   1   9 122  42   2   0
         6   0   6  44 103  38   3
         7   0   0   0  13  14   3
         8   0   0   0   0   0   0
```

Overall Statistics

```
               Accuracy : 0.5975
                 95% CI : (0.5476, 0.6459)
    No Information Rate : 0.415
    P-Value [Acc > NIR] : 1.583e-13

                  Kappa : 0.3463
 Mcnemar's Test P-Value : NA
```

Statistics by Class:

```
                     Class: 3 Class: 4 Class: 5 Class: 6 Class: 7 Class: 8
Sensitivity            0.0000   0.0000   0.7349   0.6519   0.2593    0.000
Specificity            1.0000   1.0000   0.7692   0.6240   0.9538    1.000
Pos Pred Value            NaN      NaN   0.6932   0.5309   0.4667      NaN
Neg Pred Value         0.9975   0.9625   0.8036   0.7330   0.8919    0.985
Prevalence             0.0025   0.0375   0.4150   0.3950   0.1350    0.015
Detection Rate         0.0000   0.0000   0.3050   0.2575   0.0350    0.000
Detection Prevalence   0.0000   0.0000   0.4400   0.4850   0.0750    0.000
Balanced Accuracy      0.5000   0.5000   0.7521   0.6379   0.6065    0.500
```

There is not much improvement in the accuracy here over a single tree. To see what happened, we consider the error evolution.

```
options(repr.plot.width=7, repr.plot.height=4)
plot(errorevol(adaboost.model,train))
grid()
```

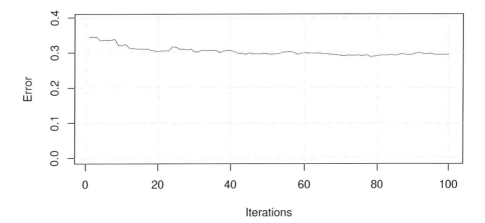

Ensemble error vs number of trees

21. Random Forests

In the random forest algorithm[1] an ensemble of decision trees is built based on the same data, similar to bagging. Recall that in bagging all of the predictor variables were considered when building each tree in the ensemble. Unlike bagging, in a random forest, only some of the features are considered. Suppose there are n features in the data set. Then each tree in the ensemble is built by randomly selecting a subset of $m \leqslant n$ features to split on. This is particularly helpful when there are a large number of features (e.g., in the hundreds). The optimal number of features used for each tree is $\sqrt{n}$. Bagging thus becomes a special case of random forests when $m = n$ is chosen. Random forests can be used to produce either a regression tree or a classification tree.

When random forests are used to produce a regression tree, the predictions of the response variable of the determined by averaging the predictions from all the trees.

When random forests are used to build a decision tree the, predictions of the classes are built by a majority vote from all of the individual trees.

The **randomForest** package[2] in R can be used to generate trees using the random forest algorithm.

Begin by loading the red wine quality data file from the UCI ML repository.

```
uci="http://archive.ics.uci.edu/"
mld="ml/machine-learning-databases/"
redcsv="wine-quality/winequality-red.csv"
wineurl=paste(uci,mld,redcsv,sep="")
wineurl
```

---

[1] Ho, TK (1995) *Random Decision Forests*. Proceedings of the 3rd International Conference on Document Analysis and Recognition (IEEE). 1:278-282; Amit, Y, Geman, D (1997) *Shape Quantization and Recognition with Randomized Trees*Neural Computation. **9**:1545-1588; Breiman, L. (2001) *Random Forests*. Machine Learning. **45**:5-32. The name *Random Forest* is a registered trademark of Leo Breiman and Andy Liaw.

[2] A. Liaw and M. Wiener (2002). *Classification and Regression by randomForest*. R News 2(3), 18–22. Full package documentation can be found at https://cran.r-project.org/web/packages/randomForest/randomForest.pdf.

```
'http://archive.ics.uci.edu/ml/machine-learning-databases/
wine-quality/winequality-red.csv'
```

```
red.wine.data=read.csv(wineurl, header = TRUE, sep = ";")
colnames(red.wine.data)
```

```
'fixed.acidity' 'volatile.acidity' 'citric.acid'
'residual.sugar' 'chlorides' 'free.sulfur.dioxide'
'total.sulfur.dioxide' 'density' 'pH' 'sulphates'
'alcohol' 'quality'
```

The last column, **'quality'**, will be the response variable. All of the other columns will be predictor variables (the features). The random forest algorithm will automatically determine which subset of features it will use for each tree in the ensemble.

We can build test and training sets. We will select 75% of the original data set as training data and put aside the remaining 25% for test data.

```
n=nrow(red.wine.data)
training.indices=sort(sample(1:n,.75*n))
train=red.wine.data[training.indices,]
train$quality = as.factor(train$quality)
test=red.wine.data[-training.indices,]
```

If this is the first time that the **randomForest** package has been used on our computer, it needs to be installed.

```
install.packages("randomForest")
```

On subsequent uses, the package will already be installed. We will only need to load the library.

```
library{RCODE}
```

With the **randomForest** function we can now build the model, which consists of an ensemble of trees. We'll use the formula method to tell **randomForest** that the column **quality** is the response variable and all the other columns are to be used as predictors.

```
rf.wine=randomForest(quality~., data=train)
```

We can print out the model, which contains a confusion matrix based on the validation runs.

```
rf.wine
```

```
Call:
 randomForest(formula = quality ~ ., data = train)
               Type of random forest: classification
                     Number of trees: 500
No. of variables tried at each split: 3

        OOB estimate of  error rate: 30.19%
Confusion matrix:
  3 4   5   6  7 8 class.error
3 0 1   7   1  0 0   1.0000000
4 1 0  24  10  0 0   1.0000000
5 0 1 427 103  2 0   0.1988743
6 0 0 107 337 19 1   0.2737069
7 0 0  11  57 73 2   0.4895105
8 0 0   0   7  8 0   1.0000000
```

The OOB estimate is the out of bag error. Because the data is sub-sampled and only some of it is used, it is possible to make a prediction using the data that was not in the bag during any given build.

We can also make a prediction and calculate a confusion matrix using data that was completely outside the training set by making a prediction with the test set.

```
library(caret)
#
# make prediction with test data
pre=predict(rf.wine,newdata=test)
#
# convert to factor form
pre.fact=as.factor(pre)
#
# print confusion matrix (quality is column 12)
confusionMatrix(pre.fact, as.factor(test[,12]))
```

The output variable err.rate in the model gives the errors in each class as well sa the out of bag error as a function of number of trees. We can make of plot of this to visualize algorithm performance.

```
Confusion Matrix and Statistics

          Reference
Prediction   3   4   5   6   7   8
         3   0   0   0   0   0   0
         4   0   0   0   0   0   0
         5   1  14 121  48   2   0
         6   0   3  26 118  29   1
         7   0   1   1   8  24   1
         8   0   0   0   0   1   1

Overall Statistics

               Accuracy : 0.66
                 95% CI : (0.6113, 0.7063)
    No Information Rate : 0.435
    P-Value [Acc > NIR] : < 2.2e-16

                  Kappa : 0.4544
 Mcnemar's Test P-Value : NA

Statistics by Class:

                     Class: 3 Class: 4 Class: 5 Class: 6 Class: 7 Class: 8
Sensitivity            0.0000    0.000   0.8176   0.6782   0.4286   0.3333
Specificity            1.0000    1.000   0.7421   0.7389   0.9680   0.9975
Pos Pred Value            NaN      NaN   0.6505   0.6667   0.6857   0.5000
Neg Pred Value         0.9975    0.955   0.8738   0.7489   0.9123   0.9950
Prevalence             0.0025    0.045   0.3700   0.4350   0.1400   0.0075
Detection Rate         0.0000    0.000   0.3025   0.2950   0.0600   0.0025
Detection Prevalence   0.0000    0.000   0.4650   0.4425   0.0875   0.0050
Balanced Accuracy      0.5000    0.500   0.7798   0.7085   0.6983   0.6654
```

```r
plot(rf.wine,col=1:7,
    main="Random Forest Error Rate for Red Wine")
options(repr.plot.width=7, repr.plot.height=4)
legend(400,.9, colnames(rf.wine$err.rate), # column $err.rate
   col=1:7,    cex=0.8,fill=1:7)
```

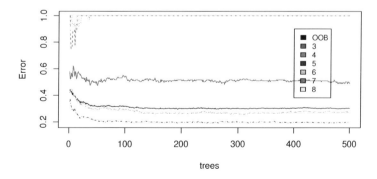

Random Forest Error Rate for Red Wine

22. K-Means Clustering

In the **k-means** clustering algorithm,[1] each cluster is associated with a *characteristic* point. This point represents the cluster, and is the centroid of the cluster. K-means is unsupervised, so it will find the cluster centers on its own. We do need to tell it how may clusters to look for, and one of the challenges of using K-means is determining the optimal number of clusters to find.

The input to k-means is the number of clusters N. Since k-means has no idea where the centroids should be, it initially makes random guesses for the characteristic centers by placing N points $\mathbf{C}_1, \ldots, \mathbf{C}_N$, at random locations.

Next, k means calculates the distance between each point in the data set $\mathbf{p}_i$, $i = 1, \ldots, n$ and each characteristic center point $\mathbf{C}_j$, $j = 1, \ldots, N$. It determines which characteristic center point is closest. Thus data point $\mathbf{p}_i$ is assigned to cluster k if $\mathbf{p}_i$ is closer to $\mathbf{C}_k$ than it is to any of the other $\mathbf{C}_j$ for $j \neq k$.

In other words, point $\mathbf{p}_i$ is assigned to the center k that *minimizes* the distance $\|\mathbf{C}_k - \mathbf{p}_i\|$ over all values of k. We say that

$$\kappa_i = \operatorname*{argmin}_{k} \|\mathbf{C}_k - \mathbf{p}_i\|$$

where κ_i is the new cluster assignment for point i. The argmin function is defined as follows: $\operatorname{argmin}(f(x))$ returns the value of x that minimizes $f(X)$.

The next step is to update the characteristic points. The cluster centers

---

[1] Kmeans was invented by Steinhaus, H. (1957). *Sur la division des corps matériels en parties.* [Bull. Acad. Polon. Sci. (in French).] 4:801-804. It was named Kmeans by MacQueen, J. (1967). *Some methods for classification and analysis of multivariate observations.* Proc. Fifth Berkeley Symp. on Math. Stat. and Prob., eds Le Cam, L. Neyman, J 1:281-297. The most common implementation is given by Lloyd, S. P. (1957). *Least squares quantization in PCM.* Technical Note, Bell Laboratories (internal publication); and Forgy, E. W. (1965). *Cluster analysis of multivariate data: efficiency vs interpretability of classifications.* Biometrics. **21**: 768-769. Lloyd's report, though it preceded Forgy, was not published publically until (1982) IEEE Trans. Inf. Theory, **28**: 128-137.

are adjusted to point to the new centroids.

$$\mathbf{C}_k = \frac{\sum_{j=1}^{n} I_{kj} \mathbf{p}_j}{\sum_{j=1}^{n} I_{kj}}$$

Here I_{ij} is an **indicator function**, so that $I_{kj} = 1$ if $\mathbf{p}_j$ is in cluster k, and zero otherwise.

The whole process is then repeated until the cluster centers don't move any more. The K-means algorithm finds the Voronoi centers (figure 22.1) of the clusters. A Voronoi diagram is a partitioning of the plane into N regions such that if a point P is in a region R with Voronoi center R', then P is closer to R' then it is to any other Voronoi center in the plane. Voronoi diagrams are named after Georgy Voronoi[2] who described them in detail in 1907. They were first mentioned in the scientific literature by René Descartes in the 17th century in a discussion on the influences of the planets.[3] In 1854, the physician John Snow analyzed outbreaks of cholera in London by comparing the locations of individual deaths to the locations of water pumps. He determined that nearly all deaths occurred in the same Voronoi cell, meaning that they were tied to a common water pump. Snow's work is one of the earliest published applications of clustering. (Snow used pipe length rather than Euclidean distance as his metric.[4]) About the same time they were also studied by Dirichlet.[5] In two dimensions individual Voronoi cells are sometimes also called **Voronoi-Dirichlet polygons** or **Thiessen polygons**.[6]

Unfortunately, there may be more than one solution, so k-means algorithm should be run multiple times with different starting points to verify that an optimal solution is reached. K-means works well with well-separated Gaussian clouds. However, guessing the correct number of clusters is crucial. Using a value of k that is too small or too big will result in grouped

---

[2] G. Voronoi, ["Nouvelles applications des paramĕtres continus à la théorie des formes quadratiques." (in French)] *Journal für die Reine und Angewandte Mathematik,* **133**:97-178 (1907)

[3] R.Descartes. [*Le Monde, ou Traite de la Lumiere* (in French)] (1644).

[4] J.Snow. *On the mode of communication of cholera.* London: John Churchill. (1855).

[5] Gustav Lejeune Dirichlet. [" Uber die Reduktion der positiven quadratischen Formen mit drei unbestimmten ganzen Zahlen." (in German)] *Journal für die Reine und Angewandte Mathematik,* **40**:209-227 (1850).

[6] Ater meteorologist After Alfred Thiessen (1872-1965) who used them in his analysis of the effects of weather on different land areas. See A. Thiessen, "Precipitation Averages for Large Areas." *Monthly Weather Review,* **39**(7):1082-1098 (1911).

Figure 22.1.: Voronoi diagram and Voronoi centers. The centers are shown as small filled diamonds. There is one cell corresponding to each center. The visualization of the Voronoi Diagram is truncated to a square, but cells around the edge either extend to infinity or edges join at some point off the boundary of the figure.

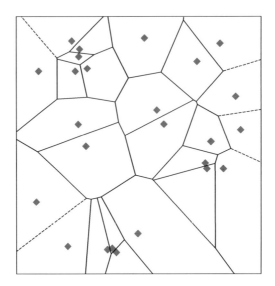

or artificially split clusters, and if the clusters are too close, they may not be well separated. Clusters that are oddly shaped, or wrap around one another, will be artificially split by K-means, and thus other algorithms would be more appropriate

K-Means on Gaussian Clouds

We'll start by illustrating how well K-means works on data that is distributed in well-separated Gaussian clouds. We will illustrated later (below) how to generate data sets like this. An example is given in the **clouds.csv** (figure 22.2.

```
clouds=read.csv("clouds.csv", header = TRUE, sep = ",")
plot(clouds$x,clouds$y,cex=0.1)
```

Note that you must replace the name of the file with the relative path the file name from your Jupyter notebook to the location of the data set. The syntax written here will only work if the data file (**clouds.csv**) is in the same folder as your Jupyter notebook (**ipynb** file).

Figure 22.2.: Two dimensional Gaussian clouds. Left: raw data. Right: same data, colored by five clusters with k-means. Centers are shown by large dots.

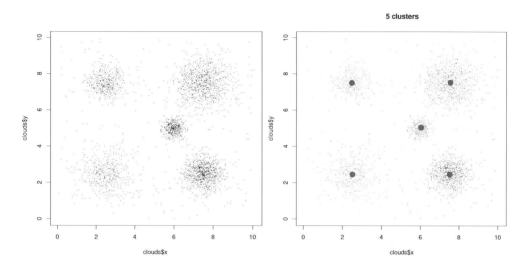

The main clustering function for k-means is called **kmeans**.[7] The standard format is:

kmeans(x, centers, options)

where **x** is the matrix of input data (columns of x, y dat) and **centers** is an integer number of centers to look for (see table 22.1).

Here we give **kmeans** the name of the data set and the number of clusters.

```
clustering=kmeans(clouds,5)
```

The output values for **kmeans** are summarized in table 22.2. We are primarily interested in plotting the cluster assignments. We will demonstrate by plotting each cluster in a different color.

To do this we will write a function that generates a random color. The **rgb(R, G, B)** function takes numbers between zero and 1 and one for the

---

[7]The algorithm used in **kmeans** is given by Hartigan, J., Wong, M. (1979) *Algorithm AS 136: A K-means clustering algorithm*. Applied Statistics. **28**:100-108.

R, G and B indices and produces a color value. The following code defines a function **random.colors(n)** that will return a list of **n** random colors.

```
random.colors=function(n){
   u=rep(0,n)
   for (j in 1:n){ u[j]=rgb(runif(1),runif(1),runif(1)) }
   u
   }
```

Table 22.1. **kmeans** Input parameters

parameter	default	description
algorithm	"Hartigan-Wong"	"Llloyd","Forgy","MacQueen", "Hartigan-Wong".
centers	-	Normally, an integer number of clusters to look form. Alternatively, can be a matrix where each row contains the initial values for the centers.
iter.max	10	maximum number of iterations .
nstart	1	Number of random sets to try.
x	-	Matrix of $x_1, x_2, ...$ data points (one column per feature; one row per point)

Table 22.2. **kmeans** Output*

betweenss	Sum of square distances between clusters. Difference between **totss** and **tot.withinss**.
centers	List of x, y values of cluster centers,
clusters	List of integers, in same order as original data pairs, giving the cluster number to which each data pair corresponds to.
iter	Number of iterations performed.
size	List of integers giving the number of points in each cluster.
tot.withinss	Total of **withinss**.
totss	Total sum of squares.
withinss	Sum of square distances within clusters. There is one value per clusters.

*Use the dollar sign operator to extract each output.

The function **runif(n)** used in the definition of **random.colors** above

184   Ch. 22. K-Means Clustering

returns **n** random uniformly distributed numbers in the interval $[0, 1]$.

Next, we want to define an array that has the same number of points as the original clustering data but each value is a color name. We will assign one color to each cluster.

First, we generate a list of five random colors with **random.colors**.

```
mycolors=random.colors(5)
```

Here we actually define the list of colors. The **rep(0,n)** function generates **n** repeats of the same number (in this case zero). We initialize the array **pointcolors** to be all zeros, even though we will override this in a minute. The purpose here is to allocate space for the array.

```
n=length(clouds$x)
pointcolors=rep(0,n)
```

Then we loop through the number of clusters, and for each cluster, we pick out all of the values that were marked in being in that cluster. If **clouds[257]**, for example, is in cluster **j**, then **pointcolors[j]** will be assigned to color **j**.

```
for (j in 1:5){
    pointcolors[clustering$cluster==j]=mycolors[j]
}
```

Note that when we do this, we loop through the clusters, and not through the points. At each iteration we pick out all the points in that cluster in one line!.

All that's left (almost!) is to plot the picture and see how well we did.

```
plot(clouds$x,clouds$y,cex=0.1,
    col=pointcolors,main=paste(5,"clusters"))
points(clustering$centers[,1],
   clustering$centers[,2], col="red",pch=19,cex=2)
```

The output of this plot is show on the right image of figure 22.2.

The Silhouette Index

The kmeans algorithm works very well when the data consists of well-separated Gaussian clouds and you know the number of clusters in advance. When you don't know the number of clusters in advanced there are several different metrics for calculating a "goodness of fit." One of these is the **silhouette index**.

To understand what this index means, suppose that $\mathbf{p}_i$ is in some cluster C. Then let

$$\mu_i(\mathbf{p}_i) = \text{mean distance from } \mathbf{p}_i \text{ to points } \mathbf{q}_j \neq \mathbf{p}_i, \forall \mathbf{q}_j \in C$$
$$\mu'_i(\mathbf{p}_i) = \text{mean distance from } \mathbf{p}_i \text{ to points } \mathbf{q}_j, \forall \mathbf{q}_j \notin C$$
$$M_i(\mathbf{p}_i) = \max_j(\mu_i, \mu'_i)$$

We can define a silhouette score for each point as

$$S_i = \frac{\mu'_i(\mathbf{p}_i) - \mu_i(\mathbf{p}_i)}{M_i(\mathbf{p}_i)}$$

Then the silhouette index is the average over all the scores,

$$S = \frac{1}{n}\sum_{i=1}^{n} S_i$$

Terms in the sum that are more negative come from points that are closer to other clusters. Thus clusterings with an average closer to positive one are better-defined.

In other words, the closer the silhouette index is to 1, the better the overall clustering.

The function `sil.score` will iterate through numerous kmeans clusterings (with different total number of clusters) find a profile of the silhouette score as a function of the number of clusters (figure 22.3). The default format is:

```
sil.score(data, nb.clus=nmin:nmax)
```

where `nb.clus` gives a range of clusters to check. One confusing aspect of this function is that the output assumes that the `nb.clus` starts with 1, so if you specify a different starting value, the extra entries will be filled with `<NA>`.

```
#install.packages("cluster")
#install.packages("bios2mds")
library(cluster)
library(bios2mds)
#
scores=sil.score(clouds,nb.clus=xclusters)
#
options(repr.plot.width=7, repr.plot.height=4)
plot(1:10, scores,type="o",xlim=c(3,10),
    xlab="Clusters", ylab="Score")
grid(lty=1,lwd=2)
```

Figure 22.3.: Silhouette score for the mixture of Gaussian clouds. The best score occurred with five clusters.

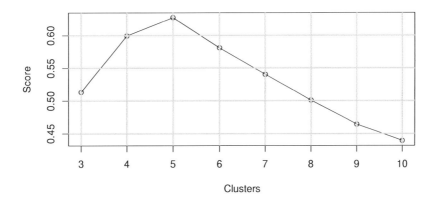

Overlapping Gaussian Clouds

In order to study kmeans we will generate some random clouds ourselves. we will use the multivariate normal random variable function **mvrnorm** from the **MASS** package.

In order to use the **mvrnorm** we need to define a covariance matrix. Normally a covariance matrix is a 2×2 matrix. We will define elliptical clouds with axes parallel to the coordinate matrices so we will use a diagonal matrix. The format of a call to **mvrnorm** is:

```
mvrnorm(n, mu, Sigma)
```

where **n** is the number of random points, **mu** gives the coordinates of the cloud center, and **Sigma** gives the covariance matrix.

Let's build five Gaussian clouds.

```
library(MASS)
sd=diag(c(.4,.2))  # covariance matrix
#
# build five gaussian clouds
#
cloud1=mvrnorm(n=1500,mu=c(1,3),Sigma=sd)
cloud2=mvrnorm(n=1500,mu=c(2.5,4),Sigma=sd)
cloud3=mvrnorm(n=1500,mu=c(4,3),Sigma=sd)
cloud4=mvrnorm(n=1500,mu=c(2,1),Sigma=sd)
cloud5=mvrnorm(n=1500,mu=c(2.5,2.5),Sigma=sd)
```

Plot the clouds on a 7×7 window (figure ??).

```
options(repr.plot.width=7, repr.plot.height=7)
plot(cloud1[,1],cloud1[,2],cex=.2, col="red",
    xlim=c(0,5),ylim=c(0,5))
points(cloud2[,1], cloud2[,2], cex=.2, col="blue")
points(cloud3[,1], cloud3[,2], cex=.2, col="orange")
points(cloud4[,1], cloud4[,2], cex=.2, col="brown")
points(cloud5[,1], cloud5[,2], cex=.2, col="green")
```

Figure 22.4.: Left: Overlapping Gaussian clouds. Right: Clustered data with 5 marked centers.

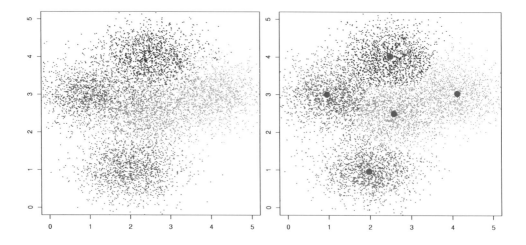

In order to do clustering, we need to combine the data into a single array.

We'll use the **rbind** function for this. This function looks for commonly labeled columns and then pastes them together one after another.

```
bigcloud=rbind(cloud1, cloud2,cloud3,cloud4,cloud5)
```

We do the clustering and plotting as we did before

```
clustering=kmeans(bigcloud,5)
n=length(bigcloud[,1])
mycolors=random.colors(5)
pointcolors=rep(0,n)
x=bigcloud[,1]
y=bigcloud[,2]
for (j in 1:k){
    pointcolors[clustering$cluster==j]=mycolors[j]
}
plot(x,y,cex=0.1,col=pointcolors,
     xlab="",ylab="")
#
#   plot centers
#   Use $ to get output from clustering
xcenters = clustering$centers[,1]
ycenters = clustering$centers[,2]
points( xcenters, ycenters, col="red",cex=2,pch=19)
```

We can see the silhouette scores still max out at five, although there is actually not much difference between three, four and five clusters in the algorithm (figure 22.5).

```
scores=sil.score(bigcloud,nb.clus=3:10)
round(scores, 2) # This prints the scores
options(repr.plot.width=7, repr.plot.height=4)
plot(1:10, scores,type="o",xlim=c(3,10),xlab="Clusters",
 ylab="Score")
grid(lty=1,lwd=2)
```

```
<NA>  <NA>  0.43  0.42  0.44  0.4  0.38  0.36  0.34  0.33
```

Kmeans for Non-Gaussian Data Sets

Here we illustrate two data sets where kmeans fails. Neither one has a Gaussian distribution. The first is composed of parallel squiggly lines. As we see from figure 22.6 this data set forms three clusters.

Figure 22.5.: Silhouette score for the overlapping Gaussian.

```
sq=read.csv("squiggles.csv", header = TRUE, sep = ",")
x=sq[,1]
y=sq[,2]
options(repr.plot.width=6, repr.plot.height=4)
plot(x,y,cex=.2,axes=FALSE,frame.plot=TRUE,xlab="",ylab="")
```

Figure 22.6.: Three parallel squiggly clusters.

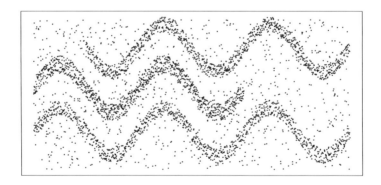

The function **kmeans** has no trouble at all doing the clustering.

```
clustering=kmeans(sq,3)
```

To avoid repeating the same block of code for assigning colors, we'll define a function that assigns colors to an array. Here we define the function `set.my.colors(clusters, kmeans.output, my.colors, XDATA)` that returns a list of colors assigned by cluster.

```
set.my.colors=function(
    clusters, kmeans.output, my.colors, XDATA){
    n=length(XDATA[,1])
    pointcolors=rep(0,n)
    ncolors = length(my.colors)
    for (j in 1:clusters){ # assing by $cluster
        pointcolors[kmeans.output$cluster==j]=my.colors[j]
    }
    pointcolors
}
```

The inputs to **set.my.colors** are: **clusters** (the number of clusters); **kmeans.output** (the exact data structure returned by **kmeans**); **my.colors** (a list of colors to draw from, in order of cluster number); and **XDATA** (the input to **kmeans**.

We'll plot the points with a small point size (**cex=0.2**) and the centers with a very large size (**cex=2**). We use **options** to change the shape of the plot window.

```
pointcolors = set.my.colors(3, clustering,
    c("red","blue", "green"), sq)
options(repr.plot.width=6, repr.plot.height=4)
plot(x,y,cex=0.2,col=pointcolors,
        xlab="",ylab="",axes=FALSE,frame.plot=TRUE)
points(clustering$centers[,1],
        clustering$centers[,2],col="black",cex=2,pch=19)
```

We can see from figure 22.7 that the result is epic failure. The problem is that the clusters found in kmeans are based on a distance from a center, and the true clusters snake around and have points will likely have points that are closer to the center of other clusters. What is really needed is a clustering method that puts together points based on *connectivity*, and not on global distance.

The second failure is the smiley. Read and plot the data:

```
smiley=read.csv("smiley.csv", header = TRUE, sep = ",")
x=smiley[,1]; y=smiley[,2]
options(repr.plot.width=4, repr.plot.height=4.5)
plot(x,y,cex=.2,axes=FALSE,frame.plot=TRUE,xlab="",ylab="")
```

Ch. 22. K-Means Clustering

Figure 22.7.: Kmeans clustering results of the three parallel squiggly clusters assuming three clusters.

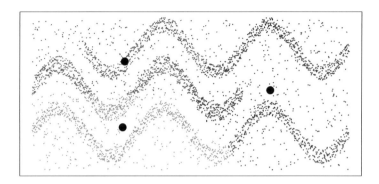

Since we can clearly count five clusters in the picture, we will ask kmeans to find them.

```
clustering=kmeans(smiley,5)
pointcolors = set.my.colors(5, clustering,
   c("red","blue", "green","pink","cyan"),  smiley)

plot(x,y,cex=0.2,col=pointcolors,
        xlab="",ylab="",axes=FALSE,frame.plot=TRUE)
points(clustering$centers[,1],
        clustering$centers[,2],col="black",cex=2,pch=19)
```

Figure 22.8.: Smiley face data (left) and clustering failure by kmeans (right) .

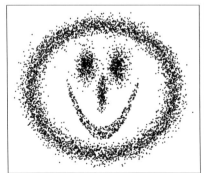 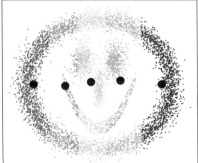

23. Hierarchical Clustering

Hierarchical clustering is another unsupervised method of finding what appear to be the natural groupings of data. It will not necessarily give the same results as other clustering methods. There are two general techniques.

- Agglomerative Clustering - Each point is considered its own cluster, and clusters are then merged iteratively to form larger clusters based on some distance measure to form larger clusters. The process continues until the entire data set is merged. The process is represented hierarchically by a binary tree, with the entire data set at the root. Thus the process is sometimes called a bottom-up approach. The most common algorithm is AGNES.

- Divisive Clustering - This is a top-down approach, with the most common algorithm being called DIANA.[1] The process starts with a single cluster (the entire data set) which is repeatedly split into smaller clusters until each cluster is a single data point.

If the trees are drawn with root node at the top, then the "natural" clusters are determined by pruning the tree with a horizontal line drawn across the plot. A silhouette measure (page 186) can be used to find the optimal number of clusters.

Agglomerative Clustering

In agglomerative clustering, the clusters are initialized so that each point comprises a single clusters. Clusters are subsequently agglomerated according to a linkage rule that depends on a distance measure $d(\mathbf{p}, \mathbf{q})$, where $\mathbf{p}$ and $\mathbf{q}$ are points. Common distance measures include:

- *Euclidean* distance $d(\mathbf{p}, \mathbf{q}) = \sqrt{\sum (p_i - q_i)^2}$. This may be best if all the clusters have Gaussian distributions.

- L_1 (Manhattan) norm $d(\mathbf{p}, \mathbf{q}) = \sum |p_i - q_i|$

---

[1] Kaufman L., Rousseeuw P.J. (1990). *Finding groups in data*. Wiley.

- L_∞ (max) norm $d(\mathbf{p}, \mathbf{q}) = \max_i |p_i - q_i|$

- *Mahalanobis* distance $d(\mathbf{p}, \mathbf{q}) = \sqrt{(\mathbf{p}-\mathbf{q})^\mathrm{T}\mathbf{S}^{-1}(\mathbf{p}-\mathbf{q})}$, where $\mathbf{S}$ is a covariance matrix. This may be best if all the clusters have multivariate normal distributions.

- *Hamming* distance for strings, i.e., the number of bits (for bit strings) or characters (for character strings) for which the two strings are different.

Initially, the distances between all points must be calculated using the chosen distance metric. The closest pair of points is joined to form a new cluster.

Subsequently, all distances must be recomputed using a linkage measure between clusters, and the distance between all pairs of clusters must be updated after each iteration. At each iteration, two clusters are joined to form a new cluster, and the between-cluster distances recalculated.[2] The process is repeated until all clusters are joined into a single tree. If A and B are clusters, some common linkage measures are:

- The *Single Link* measure: $L(A,B) = \min\{d(\mathbf{p},\mathbf{q}) | \forall \mathbf{p} \in A, \mathbf{q} \in B\}$

- The *Complete Link* measure: $L(A,B) = \max\{d(\mathbf{p},\mathbf{q}) | \forall \mathbf{p} \in A, \mathbf{q} \in B\}$

- A *group average* $L(A,B) = \dfrac{1}{|A||B|} \sum_{\mathbf{p} \in A, \mathbf{q} \in B} d(\mathbf{p},\mathbf{q})$ where $|C|$ means the number of points in cluster C. This averages the distances between all the points in either cluster.

- A *mean distance*, $L(A,B) = d\left(\dfrac{1}{|A|}\sum_{\mathbf{p} \in A}\mathbf{p}, \dfrac{1}{|B|}\sum_{\mathbf{p} \in B}\mathbf{p}\right)$. This gives the distance between the centroids of the two clusters.

- *Ward's method*: $L(A,B) = \dfrac{|A||B|}{|A|+|B|}\left[L_2\left(\dfrac{1}{|A|}\sum_{\mathbf{p} \in A}\mathbf{p} - \dfrac{1}{|B|}\sum_{\mathbf{p} \in B}\mathbf{p}\right)\right]^2,$

---

[2] The Lance-William formula provides a technique for updating these distances at each step that is more efficient that recalculating the entire matrix. Lance, G, Williams, W. (1967) *A General Theory of classificatory Sorting Strategies 1. Hierarchical Systems.* The Computer Journal. **9**:373-380.

where $L_2(\mathbf{p}) = \sqrt{\sum p_i^2}$. Ward's method is popular because it minimizes the variance.

A Worked Example

Consider the following set of data points (figure 23.1:

$$P = (1.0, 1.5)$$
$$Q = (1.0, 2.0)$$
$$R = (2.0, 2.5)$$
$$S = (4.0, 3.0)$$
$$T = (4.0, 4.0)$$

Figure 23.1.: Points {**P, Q, R, S, T**} used in the clustering example.

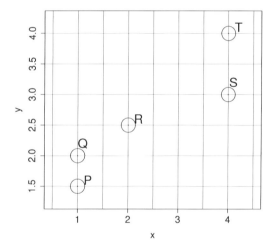

The distances between the points (using an Euclidean metric) are:

	P	Q	R	S
Q	0.50			
R	1.41	1.12		
S	3.35	3.16	2.06	
T	3.91	3.61	2.50	1.00

Ch. 23. Hierarchical Clustering

Thus **P** and **Q** are the two closest points. We can join them into a cluster. Using an averaging metric, the center of this cluster {**P**, **Q**} is at $(1, 1.75)$. The distances between the new clusters are now:

	PQ	R	S
R	1.25		
S	3.25	2.06	
T	3.75	2.50	1.00

This time, the closest pair of clusters are **S** and **T**. We form a new clusters {**S**, **T**} with a center at $(4, 3.5)$. Here are the distances between the three remaining clusters:

	ST	PQ
PQ	3.47	
R	2.24	1.25

The smallest distances is between **R** and {**P**, **Q**}. So we can form a new cluster {**R**, {**P**, **Q**}}. Since there are only two clusters left, we join them together to form the entire tree, {{**S**, **T**}, {**R**, {**P**, **Q**}}}. It is usually easier to visualize this as a *dendrogram* (figure 23.2).

Figure 23.2.: Dendrogram for the data clustered in figure 23.1.

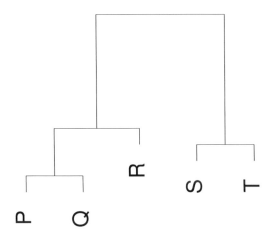

AGNES in R

Agglomerative clustering is implemented in **agnes**, in the package **cluster**.[3] We'll use the `Cars93` data set from the **MASS** package. This data set contains information about 93 different cars.

```
library(MASS)
dset=Cars93)
```

Here's what a typical row of the data set looks like (it won't fit on a single line of the page). Remember that the first number on each line tells us the row number (this is why there is a "5" every other line).

```
options(width=60)
print(dset[5,])
```

```
  Manufacturer Model    Type Min.Price Price Max.Price
5          BMW  535i Midsize      23.7    30      36.2
  MPG.city MPG.highway     AirBags DriveTrain Cylinders
5       22          30 Driver only       Rear         4
  EngineSize Horsepower  RPM Rev.per.mile Man.trans.avail
5        3.5        208 5700         2545             Yes
  Fuel.tank.capacity Passengers Length Wheelbase Width
5               21.1          4    186       109    69
  Turn.circle Rear.seat.room Luggage.room Weight   Origin
5          39             27           13   3640 non-USA
       Make
5 BMW 535i
```

For simplicity, we are going to demonstrate clustering of an all-numeric data set. The distance metric requires all the data to be numeric. Since some of the data is non-numeric we have to convert any non-numeric data into a numerical representation if we want to keep it. For this demonstration we are going to omit the non-numeric data. Before we do this, we want to keep the **Make** column, using them for the row labels.

```
row.names(dset)=dset[,"Make"]
```

We'll remove the first 2 columns, then the columns for **Type**, **AirBags**, **DriveTrain**, **Man.trans.avail**, **Make**, and **Origin**.

---

[3] Maechler, M., Rousseeuw, P., Struyf, A., Hubert, M., Hornik, K.(2018). *cluster: Cluster Analysis Basics and Extensions. R package version 2.0.7-1.* https://cran.r-project.org/package=cluster.

```
dset=dset[-c(1,2)]
dset$Type=NULL
dset$AirBags=NULL
dset$DriveTrain=NULL
dset$Man.trans.avail=NULL
dset$Make=NULL
dset$Origin=NULL
head(dset,5)
```

Here's what row 5 looks like now:

```
         Min.Price Price Max.Price MPG.city MPG.highway
BMW 535i      23.7    30      36.2       22          30
         Cylinders EngineSize Horsepower  RPM Rev.per.mile
BMW 535i         4        3.5        208 5700         2545
         Fuel.tank.capacity Passengers Length Wheelbase
BMW 535i               21.1          4    186       109
         Width Turn.circle Rear.seat.room Luggage.room
BMW 535i    69          39             27           13
         Weight
BMW 535i   3640
```

The clustering function **agnes** is in the library **cluster**.

```
library(cluster)
ag=agnes(dset)
```

It is convenient to build a **dendrogram** from the clustering results:

```
dag=as.dendrogram{ag}
```

The dendrogram is a convenient visualization of the results of clustering. It is complatible with **plot**. You will have to adjust things like the margins (**mar**), font size multiplier (**cex**), and window size based on your own configuration and the length of the label names. The default label names are the row names. This is why we copied the **make** over there. Otherwise the plot would be annotated with row number (figure 23.3).

```
options(repr.plot.width=6, repr.plot.height=16)
par(mar=c(3,4,1,8)) # margins
par(cex=.75) # font size multiplier
plot(dag,horiz=TRUE,axes=FALSE)
```

Figure 23.3.: Dendrogram for the `cars93` data using agglomerative clustering.

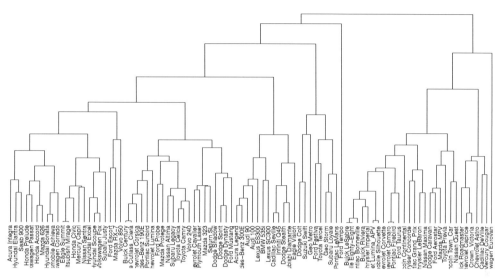

Larger trees are sometimes easier to visualize as *circular dendrograms*. Load the libraries (if necessary, install the packages).[4]

```
#install.packages("dendextend")
library(dendextend)
#install.packages("circlize")
library(circlize)
```

To get the figure to look nice you will probably need to run the code several times an manually adjust the parameters to avoid clipping labels unless all of your labels are very short. The dendrogram is shown in figure 23.4.

```
par(cex=.75)
circlize_dendrogram(dag,
    labels_track_height=.5,  # prop. of space for labels
    dend_track_height=.3)    # prop of space for tree
```

---

[4]Galili, T., et. al. (2018) *Extending 'dendrogram' Functionality in R* https://cran.r-project.org/web/packages/dendextend/. The **dendextend** includes several other functions for visualization.

Ch. 23. Hierarchical Clustering   199

Figure 23.4.: Circular dendrogram for the `Cars93` data set clustered with `agnes`.

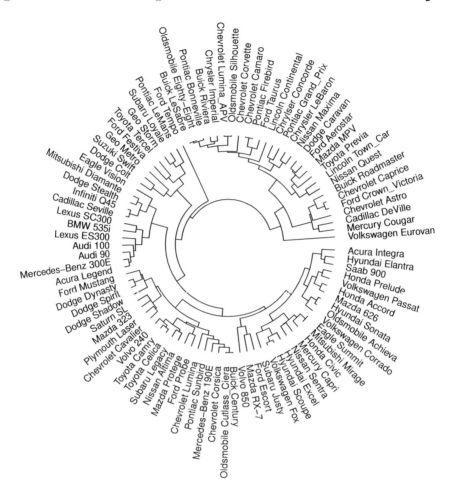

Divisive Clustering

Divisive clustering is implemented in **diana** which is also in the **cluster** package. To cluster the our modified **Cars93** dataset, convert the results to a dendrogram, and then color the branches of the dendrogram so that they show the clusters at level 3:

```
dcl=diana(dset)
ddcl = as.dendrogram(dcl)
ddcl= color_branches(ddcl, k=3)
```

The branch coloring in **color_branches** is stored in the dendrogram data structure, so it will affect visualizations made either with `plot` or with `circular_dendrogram`. colored The circular densdrogram is shown in figure 23.5.

Figure 23.5.: Circular dendrogram for the `Cars93` data set clustered with `diana`.

Comparing Hierarchical Clustering Results

The different methods of hierarchical clustering produce difference results. It is useful to have a way to compare these results. One common way of comparing results is the tanglegram. The tanglegram plots two dendrograms side by side with the leafs facing one another and then connects the matching leafs with lines. The name arises from the observation that the lines overlap or get all tangled up with one another.

```
tanglegram(cdag,ddcl,
  main_left="AGNES",
  main_right="DIANA",lab.cex=0.5)
```

Figure 23.6.: Tanglegram of the two hierarchical clusterings of the `Car93` data set.

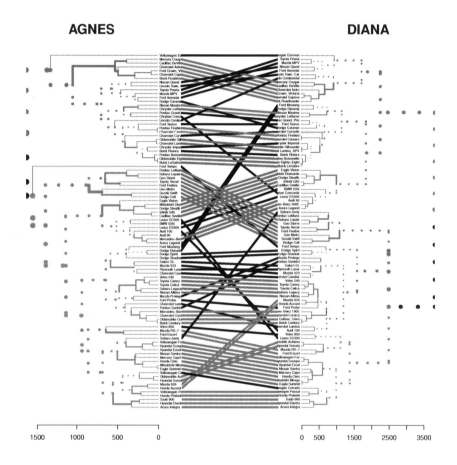

A numerical similarity measure of two cluster assignments that is often used in hierarchical clustering is the *Fowlkes-Mallows* index.[5]

---

[5] Fowlkes, E, Mallows, C. (1983) *A Method for Comparing Two Hierarchical Clusterings.* J. Amer. Stat. Assoc. **78**: 553-569. For a survey of different measures, as well as a more understandable description of the FM index, see Halkidi, M. et.

Let A and B be two different clustering assignments. Then the FM index is defined as

$$FM = \sqrt{\frac{a}{a+b}\frac{a}{a+c}}$$

where

- a = Count of pairs of points in same cluster in A and B (analagous to *true positives*.

- b = Count of pairs of points in same cluster in A but not in B (analogous to *false positives*.

- c = Count of pairs of points in same cluster in B but not in A (analogous to *false negatives*.

If the two clusterings are identical then $FM = 1$ because $b = c = 0$.

Figure 23.7.: The Fowlkes-Mallows Index as a function of number of clusters for two different heirarchical clusterings (**agnes** and **diana**) of the `Cars93` data set.

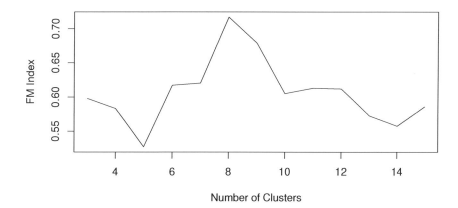

In order to calculate the FM index we need a list of cluster assignments. We can get a list of cluster assignments from hierarchical clustering tree using **cutree** in the **dendextend** library. The format is

    `cuttree(tree, k)`

---

al. (2001) *On Clustering Validation Techniques*. J. Intelligent Information Systems. **17**:107-145.

where **k** is the desired number of clusters and **tree** is a dendrogram. For example

```
cuttree(dag)
```

will return a list of the cluster assignments for the dendrogram generated above by **agnes** for the **Cars93** data.

We can write a short function to calculate the FM index for two dendrograms using the function **FM_index** from **dendextend**.

```
FM=function(A, B, j){
    FM_index(cutree(A,j),cutree(B,j))}
```

The a shorthand function to compare the dendrograms **dag** and **ddcl** is

```
FM.INDEX=function(j){FM(dag,ddcl,j)}
```

We can use this to generate a nice plot of the FM index as a function of the number of clusters.

```
options(repr.plot.width=7, repr.plot.height=4)
pdf("FM-Index-agnes-diana-car93.pdf",width=7,height=4)
plot(3:15, sapply(3:15,FM.INDEX),type="l",
     xlab="Number of Clusters",
     ylab="FM Index")
```

From the output (shown in figure 23.7) we observe that for this data set, the clusterings are most similar when we use around 8 clusters.

24. DBScan

Mathematical Description of DBSCAN

The **DBSCAN**[1] Algorithm[2] is an example of topological clustering. It determines the overall shapes of clusters by looking at points in very small neighborhoods and then connecting those neighborhoods together bit by bit. There are two input parameters: ϵ, N. We use these to define the concept of an ϵ-**neighborhood** $N_\epsilon(\mathbf{p})$ of a point,

$$N_\epsilon(\mathbf{p}) = \{\mathbf{q} | \|\mathbf{p} - \mathbf{q}\| \leq \epsilon\}$$

There are three types of points in the algorithm: **core points** (those that fall inside a cluster); **border points** (on the edge of cluster; and **noise** (points that do not fall into any cluster).

Figure 24.1.: **B** and **D** are core points; **A** and **E** are border points. **A** is directly directly density reachable from **B** but not vice-versa (using $N = 5$). **E** is density reachable from **B** but not from **A**. **A** and **E** are density connected to each other by **B**

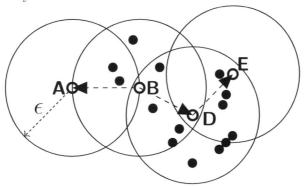

The following concepts behind DBSCAN are illustrated in figure 24.1.

- **p** is a core point if $|N_\epsilon(\mathbf{p})| \geq N$

---

[1] DBSCAN is an acronym for Density Based Spatial Clustering with (And) Noise
[2] Ester et. al (1996) *A density-based algorithm for discovering clusters in large spatial databases with noise*, KDD96, https://www.aaai.org/Papers/KDD/1996/KDD96-037.pdf The paper is easily readable by students.

- **p** is **directly density reachable** from **q** if (a) $\mathbf{p} \in N_\epsilon(q)$ and (b) $|N_\epsilon(\mathbf{q})| \geq N$.[3]

- **p** is **density reachable** from **q** if there is a sequence of points $\mathbf{q} = \mathbf{p}_1, \mathbf{p}_2, \ldots, \mathbf{p}_n = \mathbf{q}$ such that each $\mathbf{p}_{i+1}$ is directly density reachable form $\mathbf{p}_i$.

- **p** is **density connected** to **q** if there is a point **x** such that both **p** and **q** are density reachable from **x**.

- If S is a set of points, the C is a **cluster of points** in S (a) if $\mathbf{p} \in C$ and **q** is density reachable from **p**, then $\mathbf{q} \in C$; and (b) **p** is density connected to **q**.

The DBSCAN algorithm then proceeds through all the points **p** in set S. Let $C = \{\mathbf{p}\}$. Then C is expanded into the maximally density connected set containing **p**, by first examining each point in $N_\epsilon(\mathbf{p})$, and then recursively considering the ϵ neighborhood of the added points. This process is continued until all points have been clustered.

Using DBSCAN in R

There are two commonly used implementations of DBSCAN in R: the **dbscan** package[4] and the **FPC** package.[5] There is a function **dbscan** in each library, so if you happen to have both libraries loaded you should use the `library::function` format when you invoke the corresponding **dbscan** function to make sure you get the right function is used. The implementation in **FPC** came first; the documentation claims that the one in **dbscan** is faster.

We'll illustrate the DBSCAN algorithm using two contrived data sets (figure 24.2). The first one is a smiley face.

---

[3] The absolute value on the set denotes the number of points in the set.
[4] Hahsler, M. et. al. *dbscan: Density Based Clustering of Applications with Noise (DBSCAN) and Related Algorithms*. Available at https://cran.r-project.org/package=dbscan.
[5] Hennig, C. *fpc: Flexible Procedures for Clustering*. Available at https://cran.r-project.org/package=fpc.

Figure 24.2.: Clustering data for DBSCAN.

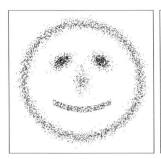
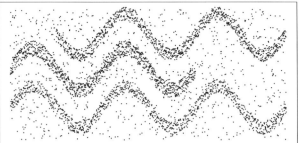

```
smiley=read.csv("smiley2.csv", header = TRUE, sep = ",")
x=smiley[,1]
y=smiley[,2]
ndata=length(x)
options(repr.plot.width=4, repr.plot.height=4.5)
plot(x,y,cex=.05,axes=FALSE,frame.plot=TRUE,xlab="",ylab="")
```

For plotting purposes, we want to define three functions. The first one, **random.colors(n)**, will return a list of **n** random colors that we can use for highlighting the different clusters.

```
random.colors=function(n){
    u=rep(0,n)
    for (j in 1:n){ u[j]=rgb(runif(1),runif(1),runif(1)) }
    u
    }
```

The second function, **set.my.colors**, will return a list of colors from a specific palette, that correspond to a list of cluster identifications. For example, if we there are a thousand points with 5 clusters, then this function will return a list of one thousand colors taken from the palette of 5 colors, where each color corresponds to a specific cluster.

```
set.my.colors=function(ncl, clust.assign, from.palette){
    ndata = length(clust.assign)
    pointcolors=rep(0,ndata)
    #ncolors = length(from.palette)
    for (j in 1:ncl){
        pointcolors[clust.assign==j]=from.palette[j]}
    pointcolors}
```

Here **ncl** is the number of clusters; **clust.assign** is a list of cluster assignments (one entry per data point); and **from.palette** is a list of color names or rgb color values that must be at least as long as the number of clusters.

Finally, we want to be able to easily produce a list of colors randomly.

```
set.random.colors=function(ncl,clust.assign){
    the.palette=random.colors(ncl)
    set.my.colors(ncl,clust.assign, the.palette)}
```

Here **set.random.colors(ncl,clust.assign)** will return a list of color assignments the same length as the number of data points, where the colors are assigned randomly.

Install the package, if necessary, and then load the library.

```
#install.packages("dbscan")
library{RCODE}
```

The main parameters to **dbscan** that need to be set are **eps**, the neighborhood size, and **minPts**. The appropriate numbers for these parameters, unfortunately, are data dependent, and you may need to tune them to get good results (table 24.1). The clustering results are plotted in figure 24.3.

```
cl=dbscan(smiley,eps=.2,minPts=10)
cl
```

```
DBSCAN clustering for 5249 objects.
Parameters: eps = 0.2, minPts = 10
The clustering contains 5 cluster(s) and 142 noise points.

   0    1    2    3    4    5
 142 3418  478  488  500  223

Available fields: cluster, eps, minPts
```

To plot the results using random color assignments,

```
clust.assign=cl$cl # extract $clusters output
pointcolors=set.random.colors(13,clust.assign)
plot(x,y,cex=0.1,col=pointcolors,
     xlab="",ylab="",axes=FALSE,frame.plot=TRUE)
```

Here we read the second data data set (squiggles).

Figure 24.3.: Clustering results for smiley face using DBSCAN.

Figure 24.4.: Cluster assignments for squiggles using DBSCAN.

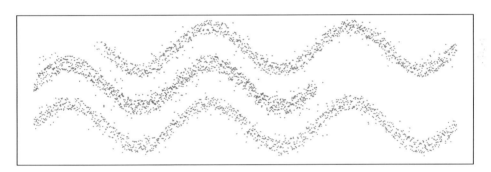

```
sq=read.csv("squiggles.csv", header = TRUE, sep = ",")
x=sq[,1]
y=sq[,2]
ndata=length(x)
```

Cluster the data. Here we used a different value **eps**, which was arrived at after some tuning.

```
dbscan(sq,eps=.075,minPts=10)
```

```
DBSCAN clustering for 4000 objects.
Parameters: eps = 0.075, minPts = 10
The clustering contains 3 cluster(s) and 467 noise points.

   0    1    2    3
 467 1177 1196 1160
```

```
Available fields: cluster, eps, minPts
```

A plot of the clustered data (figure 24.3) can be made with the same sort of code used for the smiley face, so the code is not repeated.

Table 24.1. Common **dbscan** Parameters.

dbscan(x, eps, *options* **)**
predict(object, *options* **)**

Parameter	Def.	Description
data		Data used to create original **dbscan** object. Used for predictions.
eps		Neighborhood radius.
minPts	5	Minimum number of points in a core region.
newdata		Matrix of new data points. The cluster assignments will be predicted for these points, based on an earlier fit.
object		Output of previous **dbscan** run; used only for predictions.
x		Input data (matrix).
weights		Optional list of weights for weighted clustering.

Part IV.

Other Random Topics

In this part we collect together a few methods that are often studied or mentioned in the context of machine learning but that don't really fit into the broad context of either "Classification" or "Regression."

25. Self Organizing Maps

Self organizing maps (SOM), also called **Kohonen networks**[1] or **Self Organizing Feature Maps** (SOFM) are often discussed in the context of neural networks. They were originally formulated to model certain aspects of natural brain development that contain topologically consistent representations of somatic and motor areas. An example is the *cortical sensory homunculus* (see figure 25.1),[2] which represents the location of sensory neurons in the human brain. What makes SOM's different from ANN's is that they use an area-based representation – nodes that are close together physically get updated together. If node A is near node B, then when node A gets updates, so does node B, though perhaps not as strongly.

Here is where the idea comes from. In the motor cortex, the representation for the neck is adjacent to the representation for the head. The representation for the head is next to shoulder. The representation for the shoulder is next to the representation for the upper arm. It goes on and on: upper arm, elbow, forearm, hand, fingers. In other words, its the same order as in your body. There are some discontinuities, like where the finger representation jumps to the eyes, but the basic observation is that sensory areas of the human body that are near to one another have representations (input) in adjacent areas of the brain. The same observation can be made in the motor cortex, where adjacent areas of the body have adjacent control areas in the brain.

Like a neural network, memory is represented by a collection of **nodes** or **neurons**. Nodes are defined so that it is possible to define a distance measure of some sort between them. The distance measure may be as simple as **adjacent to** or **not adjacent to** or as quantitative as a **euclidean distance**. In the first case, some kine of connectivity or positional relationship is required; in the second, positional information is required. It is also possible to define a rule in which connectivity can be derived from

---

[1] Kohonen, T. (1982). *Self-Organized Formation of Topologically Correct Feature Maps*. Biological Cybernetics. **43**:56-69.

[2] There is also a homunculus for the motor cortex. The original homunculus is attributed Penfield, W. and Boldrey, E. (1937). *Somatic Motor And Sensory Representation In The Cerebral Cortex Of Man As Studied By Electrical Stimulation*. Brain. **60**:389-443.

Figure 25.1.: The cortical sensory homunculus. A cross section of the cortex is topologically mapped to the rest of the human body.[3]

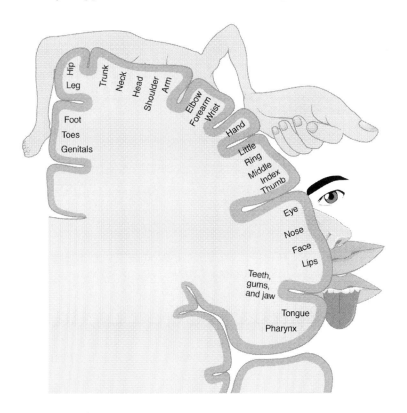

location.

In a SOM the **weights** are associated with the nodes, and not the links between nodes. Let w_i be the weight of node (location) x_i, and suppose that there is a stimulation of strength s_j at node (location) y_j. Define a *learning rate* ϵ (which may change over time, typically a small value, and typically decreasing as learning consolidates). The update to node w_i due to the stimulation y_j, is given by

$$\Delta w_i = \epsilon f(x_i, y_j)(s_j - w_i)$$

where f is some centrally peaked function that describes the probability of interaction between nodes. A simple connectivity-based near-neighbor

---

[3] Image courtesy of OpenStax College Anatomy and Physiology Web Site, https://cnx.org/contents/FPtK1zmh@6.27:KcreJ7oj@5/Central-Processing by CC 2.0 license.

probability function could take the form

$$f(x_i, y_j) = \begin{cases} 1 & \text{if } i = j \\ a & \text{if adjacent} \\ 0 & \text{otherwise} \end{cases}$$

where $0 < a < 1$. If spatial coordinates are affixed to nodes, this function could be a simple Gaussian,

$$f(x_i, y_j) = \frac{1}{\sqrt{2\pi}\sigma} e^{-(x_i-y_j)^2/(2\sigma^2)}$$

for some standard deviation σ. More commonly, a Mexican Hat Function is used:

$$f(x_i, y_j) = \left[1 - \frac{|x_i - y_j|}{K}\right] e^{-(x_i-y_j)^2/(2K)}$$

for some constant K. This function resembles a Gaussian centrally, but decreases more rapidly and adds a negative penalty further away. The Mexican Hat function becomes vanishingly small for large distances.

Kohonen Maps in R

Self organizing maps are implemented in package **kohonen**[4] For specific details the reader is referred to the reference manual at cran.

If necessary, install the package, then load the library.

```
#install.packages("kohonen")
library(kohonen)
```

We will demonstrate using the self organizing models with the wine data file at UCI. The red wine data file is at `"http://archive.ics.uci.edu/ml/machine-learning-databases/wine-quality/winequality-red.csv"`. Assume we have set the value of the variable **URL** to this string (not shown, because it won't fit a single line of the page). Then we can load the file using **read.csv**. We will convert it immediately to a matrix because

---

[4]Wehrens, R, Buydens, L. (2007) *Self- and Super-organising Maps in R: the kohonen package.* J. Stat. Softw. **21**(5). Software package available at https://cran.r-project.org/package=kohonen

the **kohonen** implementation is not compatible with data frames (the default format that data is read into), using **as.matrix**. We also want to extract the first eleven columns to use a features matrix, an use the last column for labels.

```
RED.WINE=read.csv(URL,header=TRUE,sep=";")
WINE.DATA=as.matrix(RED.WINE[,1:11])
FEATURES=head(colnames(WINE.DATA),11)
options(width=60)
print(FEATURES)
```

```
 [1] "fixed.acidity"       "volatile.acidity"
 [3] "citric.acid"         "residual.sugar"
 [5] "chlorides"           "free.sulfur.dioxide"
 [7] "total.sulfur.dioxide" "density"
 [9] "pH"                  "sulphates"
[11] "alcohol"
```

The **options(width=60)** limits the output to 60 characters per line when printing. At this point the variable **FEATURES** contains a list of the column names, and the variable **WINE.DATA** contains the feature vectors.

The original data (all 12 columns) is still in **RED.WINE**. Thats good because we still want to extract the 12th column to use as labels; the name of this column is **quality**. We'll use **classvec2classmat** which does a one-hot encoding for as well.

```
CLASSES=classvec2classmat(RED.WINE[,"quality"])
head(CLASSES)
```

3	4	5	6	7	8
0	0	1	0	0	0
0	0	1	0	0	0
0	0	1	0	0	0
0	0	0	1	0	0
0	0	1	0	0	0
0	0	1	0	0	0

The SOM impementation in **kohonen** works better if the data is centered and scaled. We can use the function **scale** to do this. The normal format is

```
scale(data, center, width)
```

Normally what **scale** does is subtract the value of **center** and then divide the resulting difference by **width**. If **data** is a matrix, then **center** and **width** should be vectors that have the same number of entries as there are columns in **data**, so that each column can be scaled differently. If either is missing, the they are replaced by the column mean

$$\mu = \frac{1}{n} \sum x_j$$

and standard deviation

$$\sigma = \sqrt{\frac{1}{n-1} \sum (x_i - \mu)^2}$$

Note that there is a different μ and σ for each column, and the values in any given column are thus scaled to their corresponding z-values,

$$z = \frac{x - \mu}{\sigma}$$

We can scale the data, and then find the SOM with **som**.

```
SOM.UNSUPERVISED=som(scaled.WINE,
   grid=somgrid(10,10,"hexagonal"))
```

where **somgrid** gives the dimensions of the desired grid, which may be either **"hexagonal"** or **"rectangular"**. For a **"rectangular"** grid, the nodes are aligned on a square grid.

There are several ways of visualizing the results. A **"codes"** plot is the default type of plot. In a **"codes"** plot, a polar are plot of wedges showing the relative significance of each feature is shown in each wedge. A polar area chart resembles a pie chart except the radius of each wedge is different. Each polar chart displays a representative feature vector. In this case, the radius of each wedge indicates the relative importance of a particular feature in a node.

```
plot(SOM.UNSUPERVISED)
```

The **codes** plot is shown in figure 25.2.

We may also be interested in knowing how many feature vectors each node in the data set actually represents. We have two ways of visualizing this. One way of visualizing this is with the **counts** plot.

Figure 25.2.: Codes plot for unsupervised SOM.

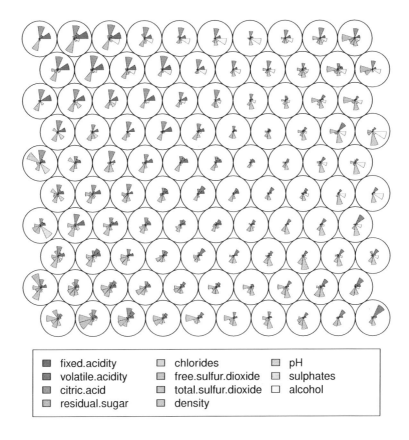

```
plot(SOM.UNSUPERVISED,type="counts",
     palette.name=topo.colors)
```

The **counts** plot is a heat map showing the relative number of vectors that fall into each category (figure 25.3).

The mapping plot draws symbols for each vector in the node (figure ??, right.)

```
plot(SOM.UNSUPERVISED, type="mapping", cex=.5)
```

Supervised clustering can be performed with the function **xyf** in the **kohonen**

Figure 25.3.: Counts (left) and mapping (right) plots for unsupervised SOM.

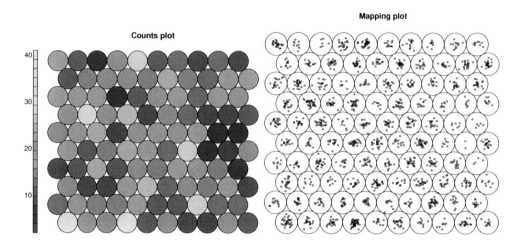

package. We train the SOM with the scaled data and the labels array **CLASSES** that we defined earlier.

```
SOM = xyf(scaled.WINE,
      CLASSES,
      grid=somgrid(7,7,"hexagonal"),
      rlen=100,
      user.weight=.5)
```

We can visualize the results with a **codes** plot (figure 25.4), a **mapping** plot (figure 25.5, left), or a **counts** plot (figure 25.5, right).

```
plot(SOM,type="codes", main="",palette.name=rainbow)
plot(SOM, type="mapping", cex=.5,main="")
plot(SOM,type="counts", palette.name=topo.colors,main="")
```

Figure 25.4.: Codes plots for supervised SOM.

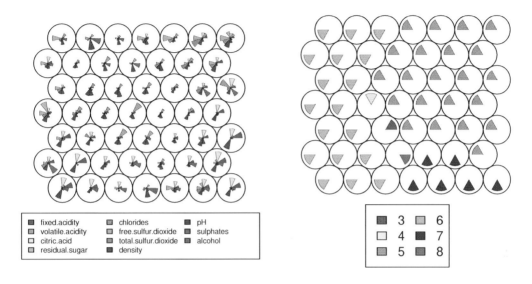

Figure 25.5.: Mapping (left) and counts (right) plots for supervised clustering with SOM.

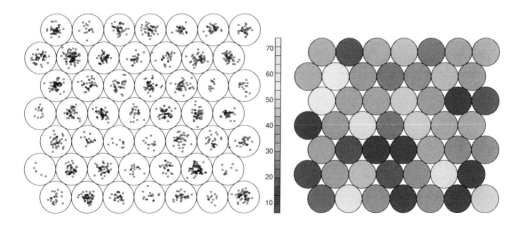

26. Hopfield Networks

Discrete Hopfield Networks

A Hopfield network is an example of a content addressable associative network. This type of network is modeled after human neural anatomy in the following senses:

- Information that is acquired at the same time is associated; for example, you can remember something by thinking about something else you were doing that day.
- Multiple and completely unrelated pieces of information can be stored in the same memory locations.
- Some portion of a memory should be able to trigger a memory response.

Hopfield networks are fully connected, i.e., each node is connected to every other node (figure 26.1). As with other neural networks, the connections

Figure 26.1.: In a Hopfield network each node is connected to every other other node. The input nodes and the output nodes are the same nodes.

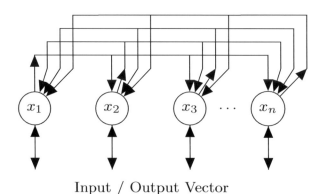

Input / Output Vector

between the nodes have weights. Denoting the connection form node i to node j as w_{ij}, the weight should increase when both x_i and x_j are active at the same time:

$$\Delta w_{ij} = \eta x_i x_j$$

for some constant η. This is called the **Hebbian learning rule**:[1] "Neurons that fire together wire together."

In a discrete Hopfield network we can store one bit (e.g., a 1 or a 0) in each neuron. A network of n neurons can be used to store an n-bit binary number. By assigning the weights properly, we can store *multiple different* n-bit binary numbers in the *same* n nodes. If $\mathbf{x}_j$ is an n-bit training vector then the learning rule gives

$$\Delta \mathbf{w} = \eta \mathbf{x}_j \mathbf{x}_j^{\mathrm{T}}$$

Summing over k training vectors and normalizing by the size k of the training set as a learning rate,[2]

$$\mathbf{w} = \frac{1}{k} \sum_{j=1}^{k} \mathbf{x}_j \mathbf{x}_j^{\mathrm{T}}$$

Component by component, we should read this as

$$w_{pq} = \frac{1}{k} \sum_{j=1}^{k} (\mathbf{x}_j)_p (\mathbf{x}_j)_q$$

where the notation $(\mathbf{x}_j)_p$ means the pth component of the training vector $\mathbf{x}_j$.

If $\mathbf{y} = (y_1, \ldots, y_n)$ is the current state of the network and $\mathbf{x} = (x_1, \ldots, x_n)$ is any other input, the learning rule will minimize the "Energy" function

$$\mathcal{E}(\mathbf{x}, \mathbf{y}) = -\frac{1}{2} \mathbf{x}^{\mathrm{T}} \mathbf{w} \mathbf{y} = -\frac{1}{2} \sum_{i,j} x_i y_j w_{ij}$$

The local minimum for a given input vector is given by the self-energy

$$\mathcal{E}(\mathbf{y}, \mathbf{y}) = -\frac{1}{2} \mathbf{y}^{\mathrm{T}} \mathbf{w} \mathbf{y} = -\frac{1}{2} \sum_{i,j} y_i y_j w_{ij}$$

This function may have multiple local minimums and we are not assured of finding a global minimum. We can write the energy function in R as follows.

---

[1] First proposed by Donald Hebb for human learning in the 1940's.
[2] Technically, we should subtract the identity matrix because the nodes do not connect to themselves. This does not affect how the network operates in the long run, except that we have non-zero diagonal on the connection matrix.

```
energy = function(x, w) {
    -0.5 * ( x %*% w %*% x)}
```

The `%*%` operator in R does matrix multiplication and handles the summation of the indices for us automatically.

To get information out of a network we use the following content addressable memory rule.

- If the net input $\mathbf{w} \cdot \mathbf{x} > 0$ to a node y_i then the output is 1.
- If the net input $\mathbf{w} \cdot \mathbf{x} < 0$ to a node y_i then the output is -1.
- If the net input $\mathbf{w} \cdot \mathbf{x} = 0$ to a node y_i the the output is y_i.

We can calculate the assignment to a single node with the following R function, where `x` is the current node value and `netin` is the net input to the node.

```
function(x, netin) {
    if (netin > 0) 1
    else if (netin < 0) -1
    else x
}
```

We are not going to give this function a name since we only have to use it in its pure form (as a *(lambda)* function). If **xvec** represents a current state vector and **net** is a vector of network totals,

```
decision = function(xvec, net) {
    mapply(
        function(x, netin) {
            if (netin > 0) 1
            else if (netin < 0) -1
            else x
        },
        xvec,
        net)
}
```

Here `decision(x, net)` returns a vector of new outputs.

Let's train a network to store three numbers. We will represent the decimal numbers 17, 3, and 42 in binary.

To convert a number to binary in R we can use the function **intToBits**. Unfortunately, this produces a vector of 32 raw byte values with the lowest bit first.

```
options(width=54)
intToBits(47)
```

```
 [1] 01 01 01 01 00 01 00 00 00 00 00 00 00 00 00 00
[17] 00 00 00 00 00 00 00 00 00 00 00 00 00 00 00 00
```

To make the problem easily understandable, we want to convert this to integers and only look at the lower eight bits. We can do that with the **as.integer** function and the **head** function.

```
head(as.integer(intToBits(47)),8)
```

```
1 1 1 1 0 1 0 0
```

We will put the bits in the correct order with the **reverse** function and then define our own **binary** function.

```
binary=function(x,n){
   rev(head(as.integer(intToBits(x)),n))}
```

Our new function **binary(x,n)** takes an integer **x** and returns an **n** bit vector of 1's and 0's.

For the neural network, we want to store the bits as -1 and 1 and not as zero and one. The conversion to this "bipolar" format is $y = 2x - 1$, where y represents the bipolar form and x is the binary form. The reverse conversion (back to binary from bipolar) is therefore $x = (y + 1)/2$.

```
bipolar=function(x){2*x-1}
frombinary=function(x){sum(c(128,64,32,16,8,4,2,1)*((x+1)/2))}
```

Our new function **bipolar(x)** takes a vector and coverts it to bipolar form. Thus **bipolar(binary(x,n)** will return an **n**-bit binary vector in bipolar form representing the integer **x**.

Out new function **frombinary(x)** takes an 8-bit bipolar binary representation and returns it to integer form.

Here is the code to put the integers 17, 3, and 42 into the columns of the matrix **data**.

```
x1=bipolar(binary(17,8))
x2=bipolar(binary(3,8))
x3=bipolar(binary(42,8))
data=matrix(c(x1, x2, x3),ncol=3)
t(data) # transpose the output for printing
```

```
-1 -1 -1  1 -1 -1 -1  1
-1 -1 -1 -1 -1 -1  1  1
-1 -1  1 -1  1 -1  1 -1
```

Here is the code to evaluate the Hopfield network on an input vector.

```
hopfield=function(input.vector, weights){
    tol = 0.01              # convergence tolerance
    max.steps = 10          # emergency evacuation

    # initialize energy function
    new.energy = energy(data[,1],weights)

    converged = FALSE
    x=input.vector
    j=0
    while ((!converged) & (j<max.steps)) {

        # calculate net input to network
        net.input = weights %*% x

        # calculate output of all nodes
        output = decision(x, net.input)

        # update energy and see if it has changed
        old.energy=new.energy
        new.energy = energy(output, weights)

        # save output as x for next iteration
        x=output

        # test for convergence
        converged = (abs(old.energy-new.energy)<tol)
        j = j+1
        }
    print(paste("converged after j=",j,
        " iterations; result=",frombinary(x)))
    }
```

Let's test it out for values near the input. First we need to initialize (train) the network. We do this be defining the weight matrix. This is given by the sum of the outer products $\mathbf{xx}^T$ of the training vectors. If we store the training vectors as the columns of a matrix $\mathbf{M}$, then

```
initialize.hopfield.weights=function(M){
    n=ncol(M)
    w=0
    for (j in 1:n){
        x=M[,j]
        weight.updates = x %*% t(x)
        w = w+ weight.updates
    }
    w/length(x)
}
```

Since we have placed the training set in the matrix **data**, we can create the weights and put them in **weight.matrix** via

```
weight.matrix=initialize.hopfield.weights(data)
```

Lets take a peek at the weights.

```
weight.matrix
```

```
 0.375  0.375  0.125  0.125  0.125  0.375 -0.125 -0.125
 0.375  0.375  0.125  0.125  0.125  0.375 -0.125 -0.125
 0.125  0.125  0.375 -0.125  0.375  0.125  0.125 -0.375
 0.125  0.125 -0.125  0.375 -0.125  0.125 -0.375  0.125
 0.125  0.125  0.375 -0.125  0.375  0.125  0.125 -0.375
 0.375  0.375  0.125  0.125  0.125  0.375 -0.125 -0.125
-0.125 -0.125  0.125 -0.375  0.125 -0.125  0.375 -0.125
-0.125 -0.125 -0.375  0.125 -0.375 -0.125 -0.125  0.375
```

Observe that the matrix is symmetric and that the diagonal is nonzero. In a true Hopfield network we would force a zero diagonal because there are not any recurrent nodes (the nodes do not directly return to themselves).

```
for (k in 15:18){
   hopfield(bipolar(binary(k,8)), weight.matrix)}
```

```
[1] "converged after j= 1  iterations;  result= 3"
[1] "converged after j= 1  iterations;  result= 17"
[1] "converged after j= 1  iterations;  result= 17"
[1] "converged after j= 1  iterations;  result= 3"
```

So for 16 and 17 it converges to 17; for 15 and 18 it converges to a random (other) stored memory value.

```
for (k in 1:4){
   hopfield(bipolar(binary(k,8)), weight.matrix)}
```

```
[1] "converged after  j= 2   iterations;   result= 1"
[1] "converged after  j= 1   iterations;   result= 3"
[1] "converged after  j= 1   iterations;   result= 3"
[1] "converged after  j= 2   iterations;   result= 1"
```

For 2 and 3 it converges to 3. For input further from 3 it converges to a random value.

```
for (k in 39:48){
   hopfield(bipolar(binary(k,8)), weight.matrix)}
```

```
[1] "converged after  j= 1   iterations;   result= 3"
[1] "converged after  j= 2   iterations;   result= 42"
[1] "converged after  j= 3   iterations;   result= 42"
[1] "converged after  j= 2   iterations;   result= 42"
[1] "converged after  j= 2   iterations;   result= 42"
[1] "converged after  j= 3   iterations;   result= 42"
[1] "converged after  j= 2   iterations;   result= 42"
[1] "converged after  j= 2   iterations;   result= 42"
[1] "converged after  j= 2   iterations;   result= 42"
[1] "converged after  j= 1   iterations;   result= 17"
```

For k=40,41,...,47, the network converges to 42, which is the nearest stored vector.

Stochastic Hopfield Networks

One problem with Hopfield networks is that the energy function may not converge to a global minimum. One way to fix this is to modify the **decision** function to make its decision based on a probability instead of an absolute yes/no choice. We can define a logistic function

$$P = \frac{1}{1 + e^{-\mathcal{E}/T}}$$

that gives this probability. This function can be derived using methods from statistical mechanics. Here $\mathcal{E}$ is the energy function defined earlier and T is a parameter called the **temperature**. When the temperature is zero, then (a) if $\mathcal{E} > 0$, then $P = 1$; and (b) if $\mathcal{E} < 0$, then $P = 0$. Thus at $T = 0$, this is a step function. For $T > 0$, this is a logistic function, with probability gradually increasing from 0 to 1 as the energy difference between two states increases.

The simulated annealing process is implemented as follows:

- Start T very large (hot).
- Let the network converge for the given temperature.
- Reduce the temperature and repeat until the temperature is very low.

Overall, the main implementation difference in reading the network is in the `decision` function.

27. Image Analysis

A digital image[1] can be represented by a two-dimensional array of numbers, where each number represents a single pixel (figure 27.1). We will therefore find it convenient to represent an image by a function $f(r, c)$ that gives the intensity or color the pixel at row r and column c. Most of image processing is based on transforming this function.

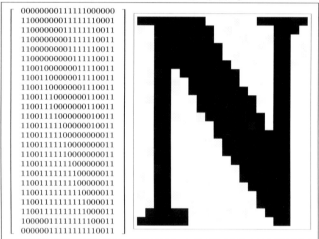

Figure 27.1.: A 25×25 pixel monochrome representation of the Northridge N as a matrix of ones and zeros (left) and the image (right).

In **monochrome images** such as(also called a black and white image, figure 27.1) the number is either a zero or a 1. In **gray-scale images** (or **gray-level images**) the number represents an intensity value, usually between 0 and 255 (for 8-bit gray-scale); sometimes we normalize this to a real number between 0 and 1. In **color images**, rather than taking on a value, $f(r, c)$ has a vector value.

Computationally, monochrome and gray-scale images are normally stored as rectangular arrays. Color images are represented in layers, e.g., in three dimensional arrays. Each layer is the same size as the full image but represents, e.g., a different color. In the **RGB** system, for example, each pixel is represented by **red** value, **blue** value, and a **green** value. The

---

[1] An excellent reference on image analysis is R.C. Gonzalez and R.E.Woods, Digital Image Processing, 3rd Edition, Prentice Hall (2008).

values range from 0 to 1 (floating point) or 0 to 255 (integral). A typical way of storing the data is to create a $h \times w \times 3$ dimensional array, where h and w are the height and width of the image in pixels.

In **CMY images** the pixels are represented by refer to **cyan**, **magenta**, and **yellow** values. The reason for having two systems is that most printers deposit pigments in each of the three CMY colors, but the light measured by most imaging systems is composed of RGB components. In fact, the two systems are equivalent, because when illuminated by white light, cyan pigment will not reflect any red; magenta pigment will not reflect any green light; and yellow pigment will not reflect any blue light. The conversion between the two systems is given by $C = 1 - R, M = 1 - G, Y = 1 - B$. Most printers also add a fourth pigment, black, which is the predominant color in printing; otherwise, equal amounts of each of the three other CMY pigments would have to be combined to produce black. The resulting system is called **CMYK**.

Physiological perception is described by a fourth representation, called **HSI** (for Hue, Saturation, Intensity). This system is sometimes useful in describing things in terms of observed colors like brown, which is easier to visualize than "a color that reflects 60% red, 40% green and 20% blue light." The conversion, where R, G, and B are normalized to [0,1], is given by the equations

$$H = \theta, \text{ if } B \leqslant G, \text{ and } 360 - \theta \text{ otherwise, where}$$
$$\theta = \cos^{-1} \frac{[(R-G) + (R-B)]/2}{\sqrt{(R-G)^2 + (R-B)(G-B)}}$$
$$S = 1 - \frac{3}{R+G+B} \times \min(R, G, B)$$
$$I = \frac{R+G+B}{3}$$

Most of what we need for image processing is in two packages in R: **imager** and **magick**. The image processing stuff is in **imager**[2] and **magick**[3] is a

---
[2] Barthelme, S. et. al. (2018) *imager: Image Processing Library Based on CImg*, available at https://cran.r-project.org/package=imager. There is an excellent getting started tutorial at https://cran.r-project.org/web/packages/imager/vignettes/gettingstarted.html.
[3] Ooms, J (2018) *majick: Advanced Graphics and Image-Processing in R*, avail-

wrapper for ImageMagick, which will allow us to read and write a variety of file formats.

We begin by loading the libraries (installing the packages if necessary).

```
# install.packages("magick")
# install.packages("imager")
library(magick)
library(imager)
```

We can load an image with the function **load.image**. You should try this with your favorite **jpg** image. Display the figure with the usual **plot** function (figure 27.2).

```
soda.pop=load.image("cokepepsi.jpg")
plot(soda.pop)
```

Figure 27.2.: Image files can be read with **load.image** and displayed on the screen with **plot**.

One thing you should notice from figure 27.2 is that the origin is at the upper left corner (unlike norm scatter plots) and that the y axis values are increasing as you go down on the page. This is because coordinates in

---

able at https://cran.r-project.org/package=magick. There is an excellent introductory tutorial at https://cran.r-project.org/web/packages/magick/vignettes/intro.html

an image are read as (row, column), so the coordinates displayed on the screen are as faithful as possible to the actual matrix representation of the image.

We can crop a picture using **imsub** (figure 27.3. You specify the range of coordinates to crop with a sequence of inequalities. A sequence of multiple inequalities may be used.

```
pepsi=imsub(soda.pop, x>165)
plot(pepsi)
```

Figure 27.3.: Image files can be cropped with **imsub**.

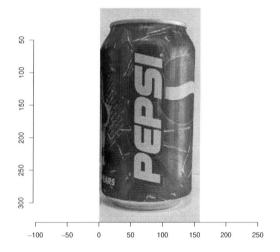

The red, blue, and green channels can be easily extracted from an image with the functions **R**, **B**, and **G** (figure 27.4). They can be combined into a gray scale image with **grayscale**.

```
red=R(soda.pop); green=G(soda.pop); blue=G(soda.pop)
par(mfcol=c(1,3))
plot(red, axes=FALSE)
plot(green, axes=FALSE)
plot(blue, axes=FALSE)
```

Figure 27.4.: Red, green, and blue levels can be extracted with **R**, **G**, and **B**.

Intensity Transformation

Intensity transformations are the basic tool of image processing. They are use to extract information from raw data and make the images more presentable. An intensity transformation **T** is an operator that converts one image representation to another:

$$g(x, y) = \mathbf{T}[f(x, y)]$$

Intensity transformations can be used, for example, to remove noise or enhance contrast. Most readers are probably quite familiar with the various intensity transformations that are built into their smart phone cameras. These allow them to enhance image quality and make their pictures look better. Such transformations are important for more than making the image pretty. They include various filtering techniques that can enhance object identification, so that machine learning techniques such as facial recognition can be applied.

Suppose we have a low contrast image, such as the upper left-hand corner of the picture of the Hollywood sign, behind the trees (the box in the top image of figure 27.5).[4] The original image was in color. We converted the image to gray level and cropped out the upper left hand corner.

```
hollywood=grayscale(load.image("hollywood.jpg"))
hollywood.ULC=imsub(hollywood, x<400, y<200)
```

The cropped image is shown on the left hand side of figure 27.6.

The distribution of image intensities across an image is given by an **image**

---

[4]Hollywood picture is public domain image from http://commons.wikimedia.org/wiki/File:Aerial_Hollywood_Sign.jpg

Figure 27.5.: Picture of the Hollywood. The top left hand corner, where we crop out a section for further analysis, is marked with a white line.

histogram. An image histogram is literally a histogram of the pixel intensities in an image. We can use the function **hist** to display a histogram (see the top image of figure 27.7).

```
options(repr.plot.width=7, repr.plot.height=3.5)
hist(hollywood.ULC, main="", xlab="Gray Level")
```

The idea of contrast stretching is to utilize all pixel levels (all gray levels) in an image. We want to spread out the histogram from a peaked, skewed, or multiply peaked distribution to a uniform distribution. We can do this using an empirical cumulative distribution function, such as **ecdf** in R. An empirical cumulative distribution function maps a distribution of maps a set of pixel weights $\{x_1, x_2, \ldots, x_n\}$ according to

$$F_n(t) = \frac{|\{x_i | x_i < t\}|}{n}$$

In R, **ecdf** returns a function that can be applied to a vector of pixels, and returns the equalized vector of pixels. Since **ecdf** returns a vector of pixels, we need to convert it back to an image file with **cimg** before we can visualize it. Imager defines the class **cimg** to store image data. We can do the equalization and plot the results in the following three lines of R code.

Ch. 27. Image Analysis

Figure 27.6.: Upper left hand corner of Hollywood sign cropped as shown in figure 27.5. Left: original cropped image. Right: after histogram equalization. The corresponding image histograms are shown in figure 27.7.

```
f=ecdf(hollywood.ULC)  # defines a function
holly=as.cimg(f(hollywood.ULC), dim=dim(hollywood.ULC))
hist(holly, main="After Equalization", xlab="Gray Level")
```

The equalized histogram is shown in the bottom of figure 27.7. The modified image is on the right hand side of figure 27.6.

Spatial Filtering

Intensity transforms may be either linear or non-linear. The most common transformation we will use is a **linear spatial filter**. Spatial filters consider all the pixels in some neighborhood of (x, y) and generate the transformed value at that location based on a fixed formula. For example, we may use a 3×3 linear spatial filter, which considers the pixels to the left, right, above, below, and diagonally adjacent. The transformed image is then given by

$$g(x,y) = \sum_{p=-1}^{1} \sum_{q=-1}^{1} k(p,q) f(x+p, y+q)$$

This is an example of a **neighborhood** filter because it only uses information in the neighborhood of pixel (x, y). For example, a filter that averages each pixel with its 8 near neighbors is

$$g(x,y) = \frac{1}{9} \sum_{p=-1}^{1} \sum_{q=-1}^{1} f(x+p, y+q)$$

Figure 27.7.: Histograms of the images shown in 27.6. Top: histogram of original cropped image. Bottom: histogram of image after equalization.

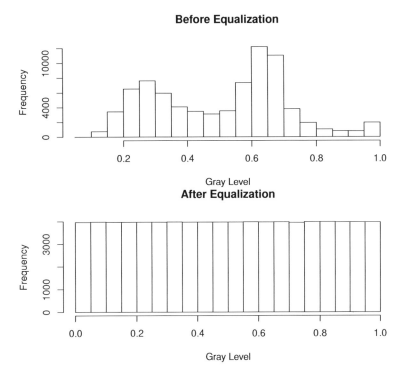

We will be primarily interested in linear filters of the form

$$K(x,y) \otimes f(x,y) = \sum_{p=-X}^{X} \sum_{q=-Y}^{Y} K(p,q) f(x+p, y+q)$$

where $K(p,q)$ is called the **filter** and the operation $\otimes$ is known as **convolution**. Conceptually, we might think of sliding a square box of dimensions $(2X+1) \times (2Y+1)$ pixels over parts of the image, and replacing the pixel in the center of the box with a weighted average that is described by the filter K. In general we also want our transformation to be normalized.

$$g(x,y) = \frac{K(x,y) \otimes f(x,y)}{\sum_{p=-X}^{X} \sum_{q=-Y}^{Y} K(p,q)}$$

Typically the filter or **kernels** can be represented by grids, where each box in the grid represents a pixel,

$$K_1 = \frac{1}{9} \times \begin{array}{|c|c|c|} \hline 1 & 1 & 1 \\ \hline 1 & 1 & 1 \\ \hline 1 & 1 & 1 \\ \hline \end{array}, \qquad K_2 = \frac{1}{16} \times \begin{array}{|c|c|c|} \hline 1 & 2 & 1 \\ \hline 2 & 4 & 2 \\ \hline 1 & 2 & 1 \\ \hline \end{array}$$

These are examples of **smoothing** filters. All of the weights in the kernel are positive and give a numerical approximation to

$$\iint K(p,q) f(x+p, y+q) dp dq$$

The first example (K_1) replaces each pixel by the average of it's intensity and the intensity of its adjacent pixels (figure 27.8); the second example (K_2) replaces the pixel with a weighted average. The idea of these filters is to remove random noise, which is usually characterized by sharp, spiky transitions. Unfortunately, edges are also characterized by these sharp transitions so smoothing tends to make the image appear blurry; this is especially pronounced for larger filters. Spatial filtering can be performed in R using either **correlate** or **convolve**. The difference between the two functions is that **convolve** replaces the addition (plus) signs before p and q in the sum above with minus signs. The two frames in figure 27.8 were generated with the following block of code.

```
G=grayscale(soda.pop)
K1 = as.cimg(matrix(1,7,7))
IG=correlate(G, K1)
plot(G, axes=FALSE, main="Unfiltered")
plot(IG, axes=FALSE, main = "Filtered")
```

It's worth noting that its actually easier to do image convolution with the **image_convolve** function in the **magick** packge; we'll demonstrate that below in the section on edge detection.

One of the purposes of smooth filtering is to remove unnecessary data that might cause confusion in feature identification. This is illustrated by the process of **thresholding**: transform every value in the image to either a zero or a 1, depending on whether it is above or below a given threshold. As shown in figure 27.9 this can remove details but still allow the identification of gross features.

Figure 27.8.: Smoothing using **correlation** using a spatial filter. Left: unfiltered image. Right: smoothed image using a 7 × 7 filter of all ones.

Unfiltered   **Filtered**

Edge Detection

Sharpening filters are the opposite of smoothing filters: their purpose is to highlight transitions in intensity, rather than hide them, primarily for the purpose of identifying edges between features in objects. Sharpening is generally obtained by differentiating the image in some way, just as smoothing is obtained by integration. Consider the definition of a derivative from calculus:

$$\frac{\partial f(x,y)}{\partial x} = \lim_{h \to 0} \frac{f(x+h,y) - f(x,y)}{h}$$

If we measure distance in terms of pixels, then the smallest non-zero distance is a single pixel; approximating $h \approx 1$ pixel will give us a numerical estimate of the derivative:

$$f_x(x,y) \approx f(x+1,y) - f(x,y)$$

This is called a **forward** approximation for the derivative at (x,y) because it uses the interval $(x, x+1)$ to approximate $f'(x,y)$. The corresponding **backward** approximation is

$$f_x(x,y) \approx f(x,y) - f(x-1,y)$$

Figure 27.9.: Result of thresholding at 0.3, 0.4, and 0.6.

because it uses the interval $(x-1, x)$. The backward approximation at x is also equal to the forward approximation at $x - 1$. Taking their difference gives an approximation for the second derivative:

$$\begin{aligned} f_{xx}(x,y) &\approx f_x(x,y) - f_x(x-1,y) \\ &\approx f(x+1,y) - 2f(x,y) + f(x-1,y) \end{aligned}$$

Figure 27.10.: Numerical derivatives of a gray-scale image. The Northridge N on the left is illustrated in gray levels, with white = 1 and black = 0. The edges taper through intermediate levels in various shades of gray. On the right is $f(x)$ along the horizontal line; $f'(x)$ (middle) and $f''(x)$ (bottom).

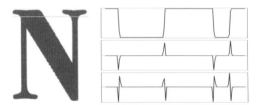

Sharp spikes occur at the transitions in the first derivative, while pairs of spikes, one up/one-down, occur in the second derivatives. The second derivative is more useful for edge detection because the passage through zero between the spikes helps to more precisely pin down its location.

Since the approximation for f_{xx} only calculates a derivative in the x−direction, only changes in intensity as we move parallel to the x−axis will be detected. Thus it will react most strongly to edges that are mainly parallel to the

$y-$ axis (vertical edges). To get horizontal edges, we should calculate f_{yy} instead of f_{xx}. This can be approximated by

$$f_{yy} \approx f(x, y+1) - 2f(x, y) + f(x, y-1)$$

To get edges in all directions, we should calculate the **Laplacian**:

$$\nabla^2 f(x,y) = f_{xx}(x,y) + f_{yy}(x,y)$$

Hence we get the following **Laplacian Filter** (see figure 27.11):

$$\nabla^2 f \approx f(x+1,y) + f(x-1,y) + f(x,y+1) + f(x,y-1) - 4f(x,y)$$

Figure 27.11.: Second derivative filters. Left to right: derivatives in the x-axis; the y-axis; along both diagonals simultaneously; along both x- and y- axes simultaneously; and Laplacian.

1	−2	1
1	−2	1
1	−2	1

1	1	1
−2	−2	−2
1	1	1

1	0	1
0	−4	0
1	0	1

0	1	0
1	−4	1
0	1	0

1	1	1
1	−8	1
1	1	1

We'll demonstrate the basic edge detection filters shown in figure 27.11 using the **image_convolve** function in **magick**. First, we will recap how to read a color file, convert to gray scale, and export the gray level image to a **jpeg** image using **imager**.

```
soda.pop=load.image("cokepepsi.jpg")
G.soda.pop=grayscale(soda.pop)
save.image(G.soda.pop,"cokepepsi-gray.jpg",quality=.9)
```

Now we want to read in the file as a **magick** image, which has a different format from an **imager** image (otherwise we could have just used it directly).

```
G=image_read("cokepepsi-gray.jpg")
```

Next, we define each of the five filters in figure 27.11. (Recall that when we assign values to a matrix in R, we do it by assigning them one column at a time.)

```
x.axis=matrix(c(1,1,1,-2,-2,-2,1,1,1), nrow=3)
y.axis=t(x.axis)
double.diagonal=matrix(c(1,0,1,0,-4,0,1,0,1),nrow=3)
both.axes=matrix(c(0,1,0,1,-4,1,0,1,0),nrow=3)
lap=matrix(c(1,1,1,1,-8,1,1,1,1),nrow=3)
```

The we can produce examples of each of the five filters as follows. Since most of the picture becomes black when we apply these filters it is useful to reverse the values of all the pixels after performing the convolution. We can do this with the **image_negate** function.

```
image_convolve(G,x.axis) %>% image_negate %>% plot()
image_convolve(G,y.axis) %>% image_negate %>% plot()
image_convolve(G,double.diagonal) %>% image_negate %>% plot()
image_convolve(G,both.axes) %>% image_negate %>% plot()
image_convolve(G,lap) %>% image_negate %>% plot()
```

The resulting plots are all shown in figure 27.12. Observe that we are using **piping** on each line of code above here. Piping is a shorthand method to avoid typing an extra line when we don't want to keep intermediate output. The **pipe** symbol in R is `%>%`. If you use linux, you are probably already familiar with piping; it works the same way in R. A pipe in R takes the output of the item on the left of the pipe operator and feeds it into the first argument of the item on the right of the pipe operator. Thus the first line above is equivalent to:

```
G1=image_convolve(G,x.axis)
G2=image_negate(G1)
plot(G2)
```

Since we don't have any reason to save the values of **G1** and **G2**, we can just do piping. If want to supply additional parameters to either of the functions, such as **plot**, that works to. The pipe knows to insert the piped input before any parameters that are included.

A variation on this is given by the Sobel Operators.[5] If we define

$$H(x,y) = S_x(x,y) \otimes f(x,y)$$
$$V(x,y) = S_y(x,y) \otimes f(x,y)$$

---

[5] I.E.Sobel. Camera Models and Machine Perception. Ph.D. Dissertation, Stanford University (1970).

Figure 27.12.: Comparison of Filters. The original image is shown on the left, followed by the using the second derivative in x, y, both diagonals, both axes, and the Laplacian. The filtered images have been color-reversed (negated) and boxes have been drawn around the edges for clarity.

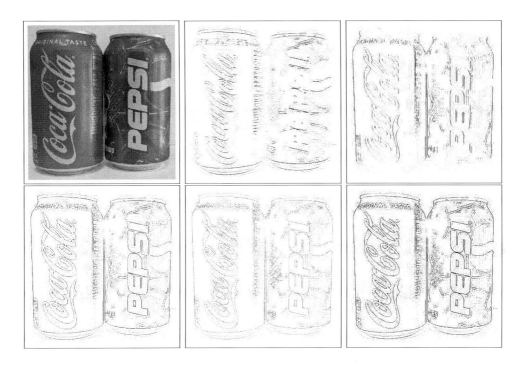

then the transformation is

$$g(x,y) = \sqrt{H(x,y)^2 + V(x,y)^2}$$

where

$$S_x = \begin{array}{|c|c|c|} \hline -1 & 2 & -1 \\ \hline 0 & 0 & 0 \\ \hline -1 & 2 & -1 \\ \hline \end{array} \qquad S_y = \begin{array}{|c|c|c|} \hline -1 & 0 & -1 \\ \hline 2 & 0 & 2 \\ \hline -1 & 0 & -1 \\ \hline \end{array}$$

An even better algorithm is given by the **Canny edge detection** model. Canny[6] observed that noise is described by a Gaussian distribution and found that the optimal Kernel for noise removal is given by

$$G(x,y) = e^{-(x^2+y^2)/(2\sigma^2)}$$

---

[6] J. Canny. A Computational Approach for Edge Detection. IEEE Transactions on Pattern Analysis Machine Intelligence. 8(6):679-698 (1986).

To use this Kernel, we first smooth the image:
$$F(x,y) = G(x,y) \otimes f(x,y)$$
and then apply the Sobel technique using F instead of f.

Morphology

So far the image analysis techniques we have examined transform images from one format to another; given one array of pixels, we convert it to another set of pixels. We generally do this to make the picture look better. To derive useful information from an an image we need to do more. We want to extract meaningful image **attributes**. Usually this involves identifying higher level shapes like **blobs**, **lines** and **spots**. To us these things have meaning: the blobs may be peoples' faces; the spots their eyes; and the lines their eyeglasses. If we can somehow recognize these features we are well on our way to identifying more complex items like chairs and pickles.

The first step, of course, is to often to make the picture look better: remove noise, smooth or sharpen the edges, and otherwise increase an image's useful information content. **Image morphology** describes the geometric structure of the image such as region boundaries and skeletons. We perform **morphological operations** to transform the image to help us identify these features. Morphological operations include erosion, dilation, opening, closing, boundary extraction, hole filling, hole filling, identification of connected components. These operations are derived from mathematical morphology.

Erosion removes components of an image by contracting boundary elements such as lines; it is useful for removing connections between elements that we think should not be there. **Dilation** increases the visibility of small features by expanding boundary elements; it is useful for bridging gaps between elements that we think really should not be there. We define the Erosion of binary image A by B as

$$\text{Erosion}(A, B) = A \ominus B = \{z | B_z \subseteq A\}$$

where B_z is the translation of image B by z; and the dilation of A by B is

$$\text{Dilation}(A, B) = A \oplus B = \bigcup_{b \in B} A_b$$

Figure 27.13.: Top: Original image (left), dilation (center), and erosion (right). Bottom row: Opening (left) and Closing (right). The borders are not part of the images.

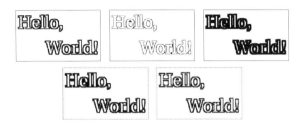

For gray scale images, the definitions are given point-wise (at point $a \in A$), by

$$(A \ominus B)(a) = \inf_{a \in B}[A(z) - B(z - a)]$$
$$(A \oplus B)(a) = \sup_{b \in B}[A(z) - B(z - a)]$$

Erosion and dilation operations can be performed by the functions `erode` and `dilate` in `imager`.

Opening and **closing** are global generalizations of erosion and dilation. Opening smooths the contours of an image by removing protrusions and narrow connections:
$$A \circ B = (A \ominus B) \oplus B$$
Closing fuses narrow breaks and eliminates small holes,
$$A \bullet B = (A \oplus B) \ominus B$$

Opening and closing can be performed in `imager` by `mopening` and `mclosing`.

Boundary extraction is similar to edge detection. It finds the boundaries of regions in an image. The operations

$$T(A) = (A - (A \ominus B))^c$$
$$S(A) = (A - (A \oplus B))$$

can both do this. Here I^C is the complement (i.e., negation) of image I,

$$I(x,y)^c = 1 - I(x,y)$$

and image subtraction is defined point-wise by subtracting pixels:

$$(A - B)(x, y) = A(x, y) - B(x, y)$$

Sometimes there are holes in our image that shouldn't be there. A **hole** is defined as a background region surrounded by a connected border of foreground pixels. **Hole filling**, also called **image completion**, is defined by the following transformation. Given an **indicator** point inside a hole, fill the entire hole, e.g., set all the pixels to 1. Let I be the image and $P_0(x, y)$ be an array that only has 1's at the indicator points, and zeros elsewhere. Then define the sequence

$$P_k = (P_{k-1} \oplus B) \bigcap I^c, \text{ where } B = \begin{pmatrix} 0 & 1 & 0 \\ 1 & 1 & 1 \\ 0 & 1 & 0 \end{pmatrix}$$

Holes are filled when the fixed point is reached, i.e., $P_k = P_{k-1}$. Hole filling can be performed in **imager** with the functions `clean` and `fill`.

Connected Components. In a binary image, two adjacent pixels are said to be **connected** if they are both 1. A pixel is said to be **connected** to the pixel next to it if the two pixels are connected. **Adjacency** can be defined in several ways. The most common are: (a) they are **4-adjacent** if they are above, below, left, or right of one another; and (b) they are **8-adjacent** if they 4-adjacent or diagonally touching. Given any two pixels, they are considered connected if there is some path between them consisting of adjacent connected pixels. Given any pixel, its **connected component** is the set of all pixels that are connected to it. The extraction of connected components is similar to the process of hole filling. Here we define an array of indicator pixels $S_0(x, y)$ where there is one pixel turned on in each connected component. Then we do fixed point iteration on

$$S_k = (S_{k-1} \oplus B) \bigcap I \tag{27.1}$$

Here B is a matrix that defines adjacency: for 4-connectedness it is like the matrix in equation **??**; for 8-connectedness it would be a matrix of all-ones.

Figure 27.14.: Left: Cross-section of the shoot apical meristem of Arabidopsis Thaliana obtained via confocal microscopy. Right: Segmentation after blurring and extraction of the morphological boundary by thresholding at $T = 0.84$.

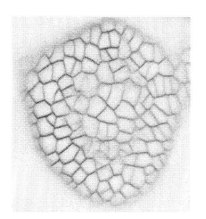
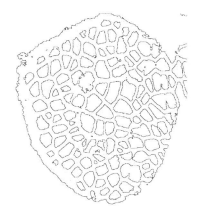

Segmentation

The next step is to divide the picture into subregions that are somehow distinct from one another. This process of separating out the different regions is known as image **segmentation**. Usually we do some sort of edge detection before we do segmentation, e.g., with a Laplacian, Sobel, or Canny Filter. If the image is noisy we may actually want to blur it first

The simplest segmentation method is **thresholding**. Every pixel in the image is converted into either 1's or 0's depending on whether or not it is above a given threshold. This works well if the items we are attempting to identify are in high contrast with the background environment. Each connected component then defines a particular segment of the image.

One very popular segmentation technique is the **Watershed Transform**. The Watershed transform is implemented in the **EBImage** package in the **Bioconductor** repository.[7] The concept is based on treating the array of pixels as a topological map. Bright pixels are at the tops of mountains and

---

[7] Pau, G. (2010) (2010). *EBImage - an R package for image processing with applications to cellular phenotypes.* Bioinformatics **26**:979-981. Details on the R package at given at http://bioconductor.org/packages/release/bioc/html/EBImage.html. Note that Bioconductor packages should be installed using **bioclite**.

dark pixels are at the bottoms of valleys (or vice-versa). If we drop water on any given pixel it will either (a) stay where it is, if it is at the bottom of a valley; (b) roll down the hill in a unique direction (given by the direction of steepest decrease); or (c) may move in one of two directions, if it is on a ridge line. We define a **watershed** as the set of all points that will fall to a particular minimum (there may be several adjacent pixels at the bottom of the basis that satisfy this minimum). The lines at the crests of the hills are called **watershed lines**. The idea is to punch holes at the bottom of each basin and slowly pump water in from the bottom, flooding the image. As the water rises, each basin slowly fills, and will eventually overflow from one watershed into another watershed. When this happens we construct an infinitely high dam to prevent water from flowing between the regions. We continue to pump water in, extending the lengths of our dams as necessary. At the end of the process, the collection of dams gives the watershed lines of the image. The problem with watershed is that the domains need to be seeded – somebody or some algorithm needs to determine where to "punch the holes" that are used to start filling in the individual domains.

28. Afterward

In this book I have not attempted to cover every subject in machine learning. Hopefully there has been enough material here to get you interested. New and interesting topics and applications are being developed (it seems like) every day. Here are a few subjects you might want to explore on your own:

- Natural language processing (NLP) and understanding human language; applications include understanding people, the media, and literature.

- Sentiment analysis: an application of NLP and computational linguistics that can be used to explore emotional and psychological attributes of text.

- Text processing algorithms such as bag of words, text classification, and spam filtering

- Connectivity and near neighbor algorithms: these algorithms study graphs and look at questions like the "six degrees of separation" or the optimal number of friends on a social network.

- Convolutional neural networks - these are deeply layered neural networks that are highly interconnected.

- Generative adversarial networks (GAN) - unsupervised neural networks that are trained in a game-like process

- Recurrent deep neural networks - networks that are not just feedforward (an example is the Hopfield network) are not that well understood, and truly deep networks have not really been explored.

The simple input/simple output approaches shown in this book will only take you so far. Keep in mind that data science and machine learning is more than a black box approach to data analysis. We have only touched the very surface. Think if it this way (by analogy): by reading this book, you now know how to use a calculator, but if you really want to understand numbers you need to learn how to do long division. There is a lot of math and a lot of code behind each method. You will need to learn how to tweak these methods and algorithms if you want to get better results. Here are a few pointers:

- Learn the math. There is no substitute for a solid foundation in linear algebra and basic statistics.

- Read the documentation on any function you are using. Figure out when it works and when it does not work.

- Use the best language (in your opinion) for each task. It is common for a data scientist to split a single project into steps that are performed in SQL, python, R, and so forth, because each language has features that make particular tasks easier.

- Always do some data exploration first. Don't blindly send it to the black box. Expect to spend 90% of your time cleaning up dirty csv files.

- Remember that a lot of useful numerical data analysis can be done using methods and techniques that are not considered part of the "machine learning" or "data science" genres. Examples include Fourier analysis, simulation, descriptive and predictive statistics, and numerical analysis. For any particular data set or scientific problem you are studying it is worth considering *all* of the methods that are available to you.

- Explore the literature, including the various dynamic online forums. Don't be afraid to join a forum and ask a question.

Now go forth and hack.

Part V.
Stuff in the Back of the Book

Extra cool stuff in the back of the book.

A. Review of Linear Algebra

Vectors

This appendix provides a brief of review of some basic concepts from linear algebra.[1] We begin with vectors.

> **Definition A.1. Vector**
>
> A **vector v** is an object with a **magnitude** and **direction**.

A vector, of and by itself, is not fixed to any particular location in space. There is a common visualization (that we will use) in which vectors are represented by arrows in $\mathbb{R}^n$. If we place the tail of the arrow at the origin then its point will be at some point **P**. We can represent the point by its coordinates. In 3 dimensions, we sometimes write the coordinates of the point is

$$\mathbf{P} = (x, y, z)$$

Vectors are typically represented by the coordinates of the point we have just described. In this case we will usually write the vector in its **column** form

$$\mathbf{v} = \begin{bmatrix} x \\ y \\ z \end{bmatrix}$$

We will sometimes write the vector in its row form as

$$\mathbf{v} = \begin{bmatrix} x & y & z \end{bmatrix}$$

Because there is a one-to-one relationship between the set of all vectors in $\mathbb{R}^n$ and the points in $\mathbb{R}^n$, the row and column representations are often used interchangeably, although (technically) they represent different objects (one is a row matrix, the other is a column matrix). You will have to determine from the context which one is intended.

---

[1] The material in this appendix was drawn from the author's text *Scientific Computation, 3rd ed.*, App. B, and Ch. 43.

In higher dimensions we will replace the symbols $x, y, z, \ldots$ with $x_1, x_2, \ldots, x_n$, and write a vector as

$$\mathbf{v} = \begin{bmatrix} v_1 & v_2 & \cdots & v_n \end{bmatrix}$$

To save space, the column form is often written as $\mathbf{v} = \begin{bmatrix} v_1 & v_2 & \cdots & v_n \end{bmatrix}^{\mathrm{T}}$ where the T is read as **transpose**. The word **transpose** means flip all the rows and columns.

The magnitude of **v** is defined as the Euclidean length of the tip-to-tail line segment that corresponds to the vector. In other words, if you place the tail of the vector at the origin, the magnitude of the **v** is the distance from the origin to the head of the vector.

> **Definition A.2. Vector Magnitude**
>
> The **magnitude, absolute value** or **length** of a vector $\mathbf{v} = [v_1, \ldots, v_n]^{\mathrm{T}}$ is given by the positive square root
>
> $$v = |\mathbf{v}| = \sqrt{v_1^2 + v_2^2 + \cdots + v_n^2}$$

We could define a function to calculate the vector magnitude in R. This will work with any length vector:

```
vnorm=function(x){sqrt(sum(x^2))}
v=c(3,7,-12)
vnorm(v)
```

14.2126704035519

We can also use the built in **norm** function which is actually a matrix norm. To use it you must specify the type of norm. For the Euclidean 2-norm you should specify **type="s"** for spectral norm.

```
norm(v,type="2")
```

14.2126704035519

We can visualize the angle between two vectors by placing their tails at the same point in space. Since the two arrows will fall in a single plane (or $n-1$ dimensional subspace), we define the angle between the vectors as the angle between the arrows in that plane.

> **Definition A.3. Angle Between Vectors**
>
> If we place **v** and **w** at the origin so that their tips are at points **P** and **Q**, respectively, then the angle between **v** and **w** is defined to be the angle between the line segments $\overline{OP}$ and $\overline{OQ}$.

We will define angles in higher dimensions in terms of the dot product.

> **Definition A.4. Dot Product**
>
> Let **v** and **w** be vectors. Their **dot product** is given by
> $$\mathbf{v} \cdot \mathbf{w} = v_1 w_1 + v_2 w_2 + v_3 w_3 + \cdots + v_n w_n$$

The dot product in R can be found by either summing over the product of components

```
v=c(1,2,3,4,5)
w=c(7,-9,10,9,16)
sum(v*w)
```

135

or by using the matrix multiplication operator **v %*% w**,

```
v %*% w
```

135

The angle θ between two vectors is defined in terms of the normal trigonometric cosine function.
$$\mathbf{v} \cdot \mathbf{w} = |\mathbf{v}||\mathbf{w}| \cos \theta$$

The sum of two vectors **v** and **w** is a vector **x** given by
$$\mathbf{x} = \begin{bmatrix} v_1 + w_1 & v_2 + w_2 & \cdots & v_n + w_n \end{bmatrix}^{\mathrm{T}}$$

If $k \in \mathbb{R}$, then the **scalar product** of k and a vector **v** is

$$k\mathbf{v} = \begin{bmatrix} kv_1 & kv_2 & \cdots & kv_n \end{bmatrix}^\mathrm{T}$$

A set is said to be **closed** under an operation if you can perform that operation on any combination of set elements and still get another element of the original set. Since any sum of vectors is a vector, the set of all vectors is said to be closed under vector addition. Similarly, since scalar multiplication always produces another vector, the set of all vectors is closed under scalar multiplication. This is called the closure property of vector addition and scalar multiplication.

> **Theorem A.1. Closure**
>
> The set of all vectors in $\mathbb{R}^3$ is closed under vector addition and scalar multiplication.

In three dimensions, the set of **standard basis vectors** is written as

$$\mathbf{i} = \begin{bmatrix} 1 & 0 & 0 \end{bmatrix}^\mathrm{T}$$
$$\mathbf{j} = \begin{bmatrix} 0 & 1 & 0 \end{bmatrix}^\mathrm{T}$$
$$\mathbf{k} = \begin{bmatrix} 0 & 0 & 1 \end{bmatrix}^\mathrm{T}$$

We can write any vector **v** as a combination of basis vectors:

$$\mathbf{v} = \begin{bmatrix} v_1 \\ v_2 \\ v_3 \end{bmatrix} = \begin{bmatrix} v_1 \\ 0 \\ 0 \end{bmatrix} + \begin{bmatrix} 0 \\ v_2 \\ 0 \end{bmatrix} + \begin{bmatrix} 0 \\ 0 \\ v_3 \end{bmatrix} = v_1\mathbf{i} + v_2\mathbf{j} + v_3\mathbf{k}$$

In higher dimensions we typically use the notation $\mathbf{e}_1, \mathbf{e}_2, \ldots$ for the basis vectors, where

$$\mathbf{e}_1 = [1, 0, 0, \ldots, 0]$$
$$\mathbf{e}_2 = [0, 1, 0, \ldots, 0]$$
$$\vdots$$
$$\mathbf{e}_n = [0, 0, 0, \ldots, 1]$$

With this notation an n-dimensional vector can be written as

$$\mathbf{v} = v_1\mathbf{e}_1 + v_2\mathbf{e}_2 + v_3\mathbf{e}_3 + \cdots + v_n\mathbf{e}_n$$

Appendix A. Review of Linear Algebra

Definition A.5. Linear Dependence

The vectors $\mathbf{v}_1, \mathbf{v}_2, \ldots, \mathbf{v}_n$ are said to be **linearly dependent** if there exist numbers $a_1, a_2, \ldots, a_n$, not all zero, such that

$$a_1\mathbf{v}_1 + a_2\mathbf{v}_2 + \cdots + a_n\mathbf{v}_n = 0$$

If no such numbers exist the vectors are said to be **linearly independent**.

Matrices

A matrix is a bunch of numbers written as a grid in a box. Individual elements are referenced by row and column, where the row index comes first and the column index comes second.

Definition A.6. Matrix

An $m \times n$ (or m by n) **matrix** A is a rectangular array of number with m rows and n columns

If we denote the number in the ith row and j^{th} column as a_{ij} then we can write a matrix $\mathbf{A}$ as

$$\mathbf{A} = \begin{bmatrix} a_{11} & a_{12} & \cdots & a_{1n} \\ a_{21} & a_{22} & & a_{2n} \\ \vdots & & & \vdots \\ a_{m1} & a_{m2} & \cdots & a_{mn} \end{bmatrix}$$

We will sometimes denote the matrix $\mathbf{A}$ by $[a_{ij}]$.

Definition A.7. Transpose

The **transpose** $\mathbf{A}^{\text{T}}$ of the matrix $\mathbf{A}$ is the matrix obtained by interchanging the row and column indices.

If **A** is as given above, then

$$\mathbf{A}^{\mathrm{T}} = \begin{bmatrix} a_{11} & a_{21} & \cdots & a_{m1} \\ a_{12} & a_{22} & & a_{m2} \\ \vdots & & & \vdots \\ a_{1n} & a_{2n} & \cdots & a_{mn} \end{bmatrix}$$

Note that $\left(\mathbf{A}^{\mathrm{T}}\right)_{ij} = \mathbf{A}_{ji}$

The transpose of an $m \times n$ matrix is an $n \times m$ matrix.

In R, the matrix transpose is obtained with the **t(M)** function.

```
M=matrix(c(1,0,3,4,5,0,0,3,1), nrow=3)
M
```

1	4	0
0	5	3
3	0	1

```
t(M)
```

$$\mathbf{A} = \begin{bmatrix} 1 & 0 & 3 \\ 4 & 5 & 0 \\ 0 & 3 & 1 \end{bmatrix}$$

1	0	3
4	5	0
0	3	1

Matrix Addition is defined between two matrices of the same size, by adding corresponding elements. Matrices that have different sizes cannot be added.

$$\begin{bmatrix} a_{11} & a_{12} & \cdots \\ a_{21} & a_{22} & \cdots \\ \vdots & & \end{bmatrix} + \begin{bmatrix} b_{11} & b_{12} & \cdots \\ b_{21} & b_{22} & \cdots \\ \vdots & & \end{bmatrix} = \begin{bmatrix} a_{11}+b_{11} & b_{22}+b_{12} & \cdots \\ a_{21}+b_{21} & a_{22}+b_{22} & \cdots \\ \vdots & & \end{bmatrix}$$

A **square matrix** is any matrix with the same number of rows as columns. The **order** of the square matrix is the number of rows (or columns).

A **submatrix** of **A** is the matrix **A** with one (or more) rows and/or one (or more) columns deleted.

The **determinant** $\det \mathbf{A}$ **of a square** $n \times n$ **matrix A** is calculated as follows.

1. If $n = 1$ then $\mathbf{A} = [a]$ and $\det \mathbf{A} = a$.

2. If $n \geq 2$ then
$$\det \mathbf{A} = \sum_{i=1}^{n} a_{ki}(-1)^{i+k}\det(\mathbf{A}'_{ik})$$
where $\mathbf{A}'_{ik}$ is the submatrix of A with the ith row and k^{th} column deleted.

The choice of which k does not matter because the result will be the same.

The determinant is sometimes written as
$$\det \mathbf{A} = \begin{vmatrix} a_{11} & a_{12} & \cdots \\ a_{21} & a_{22} & \cdots \\ \vdots & & \end{vmatrix}$$

In particular, for a 2×2 matrix,
$$\begin{vmatrix} a & b \\ c & d \end{vmatrix} = ad - bc$$

and for a 3×3 matrix,
$$\begin{vmatrix} A & B & C \\ D & E & F \\ G & H & I \end{vmatrix} = A\begin{vmatrix} E & F \\ H & I \end{vmatrix} - B\begin{vmatrix} D & F \\ G & I \end{vmatrix} + C\begin{vmatrix} D & E \\ G & H \end{vmatrix}$$

As an example, consider
$$\mathbf{A} = \begin{bmatrix} 1 & 0 & 3 \\ 4 & 5 & 0 \\ 0 & 3 & 1 \end{bmatrix}$$

Then
$$\det \mathbf{A} = \begin{vmatrix} 1 & 0 & 3 \\ 4 & 5 & 0 \\ 0 & 3 & 1 \end{vmatrix} = (1)\begin{vmatrix} 5 & 0 \\ 3 & 1 \end{vmatrix} - 0\begin{vmatrix} 4 & 0 \\ 0 & 1 \end{vmatrix} + 3\begin{vmatrix} 4 & 5 \\ 0 & 3 \end{vmatrix} = 41$$

In R, we can use the **det** function.

```
M=matrix(c(1,0,3,4,5,0,0,3,1), nrow=3)
R
```

```
1    0    3
4    5    0
0    3    1
```

```
det(M)
```

```
41
```

A important property of a matrix is whether or not it is singular. A matrix is singular if and only if its determinant is zero. The matrix has an inverse (discussed below) if and only if it is nonsingular.

Definition A.8. Singular Matrix

A square matrix **A** is said to be **singular** if $\det \mathbf{A} = 0$, and **non-singular** if $\det \mathbf{A} \neq 0$.

Theorem A.2.

The n columns (or rows) of an $n \times n$ square matrix A are linearly independent if and only if $\det \mathbf{A} \neq 0$.

Definition A.9. Matrix Multiplication

The product of an $m \times r$ matrix **A** and an $r \times n$ matrix **B** is the $m \times n$ matrix **AB** with elements

$$\mathbf{AB}_{ij} = \sum_{k=1}^{r} a_{ik} b_{kj} = \sum_{k=1}^{r} \mathbf{a}_i \cdot \mathbf{b}_j$$

where $\mathbf{a}_i$ is the ith row of **A** and $\mathbf{b}_j$ is the jth column of **B**.

For example,

$$\begin{bmatrix} 1 & 2 & 3 \\ 4 & 5 & 6 \end{bmatrix} \begin{bmatrix} 8 & 9 \\ 10 & 11 \\ 12 & 13 \end{bmatrix} = \begin{bmatrix} (1,2,3) \cdot (8,10,12) & (1,2,3) \cdot (9,11,13) \\ (4,5,6) \cdot (8,10,12) & (4,5,6) \cdot (9,11,13) \end{bmatrix}$$

$$= \begin{bmatrix} 64 & 70 \\ 156 & 169 \end{bmatrix}$$

Here is the same calculation in R:

```
A=t(matrix(c(1,2,3,4,5,6),nrow=3))
B=t(matrix(c(8,9,10,11,12,13),nrow=2))
print(A)
print(B)
C=A %*% B
print(C)
```

```
     [,1] [,2] [,3]
[1,]    1    2    3
[2,]    4    5    6
     [,1] [,2]
[1,]    8    9
[2,]   10   11
[3,]   12   13
     [,1] [,2]
[1,]   64   70
[2,]  154  169
```

When we multiply matrices the dimensions must match. In the example, we multiplied a $[2 \times 3] \times [3 \times 2]$ matrix. The middle dimension here matched (3) so we are able to multiply. If the middle dimensions do not match, we cannot multiple. Thus, for example, we cannot multiply a $[2 \times 3]$ matrix by itself because $[2 \times 3] \times [2 \times 3]$ does not nave matching middle dimensions.

Theorem A.3. Determinant of Product

If **A** and **B** are square matrices then

$$\det \mathbf{AB} = (\det \mathbf{A})(\det \mathbf{B})$$

Definition A.10. Main Diagonal and Diagonal Matrices

The **main diagonal** of a square matrix **A** is the sequence of numbers $a_{11}, a_{22}, \ldots, a_{nn}$. A square matrix is said to be **a diagonal matrix** if its only non-zero elements lie on its main diagonal.

An identity matrix of degree n is a special $n \times n$ diagonal matrix that has 1's on its main diagonal and zeros everywhere else.

Definition A.11. Identity Matrix

The $n \times n$ **identity matrix** is the matrix **I** that satisfies the equation

$$\mathbf{AI} = \mathbf{IA} = \mathbf{A}$$

for all $n \times n$ matrices **A**

The identity matrix is a diagonal matrix with 1's in the diagonal and zero's everywhere else.

Definition A.12. Matrix Inverse

A square matrix **A** is said to be **invertible** if there exists a matrix $\mathbf{A}^{-1}$, called the **inverse** of **A**, such that

$$\mathbf{AA}^{-1} = \mathbf{A}^{-1}\mathbf{A} = \mathbf{I}$$

Theorem A.4. Requirement for Invertibility

A square matrix is invertible if and only if it is nonsingular.

Definition A.13. Matrix Cofactor

Let **A** be a square $n \times n$ matrix. Then the ij **cofactor**, is

$$\operatorname{cof} a_{ij} = (-1)^{i+j} \det \mathbf{M}_{ij}$$

where $\mathbf{M}_{ij}$ is the submatrix of **A** with row i and column j removed.

For example, let us find the cofactor of a_{12} in

$$A = \begin{bmatrix} 1 & 2 & 3 \\ 4 & 5 & 6 \\ 7 & 8 & 9 \end{bmatrix}$$

The required cofactor is

$$\operatorname{cof} a_{12} = (-1)^{1+2} \begin{vmatrix} 4 & 6 \\ 7 & 9 \end{vmatrix} = (-1)(36 - 42) = 6$$

Definition A.14. Classical Adjoint Matrix

If **A** is an $n \times n$ matrix then its **classical adjoint**, denoted by adj **A**, is the transpose of the matrix that results when every element of **A** is replaced by its cofactor.

$$\operatorname{adj} \mathbf{A} = [\operatorname{cof} a_{ij}]^{\mathrm{T}} = (\operatorname{cof} \mathbf{A})^{\mathrm{T}}$$

As an example, let us find the classical adjoint of

$$\mathbf{A} = \begin{bmatrix} 1 & 0 & 3 \\ 4 & 5 & 0 \\ 0 & 3 & 1 \end{bmatrix}$$

The matrix of cofactors is

$$\operatorname{cof} \mathbf{A} = \begin{bmatrix} (1)\begin{vmatrix} 5 & 0 \\ 3 & 1 \end{vmatrix} & (-1)\begin{vmatrix} 4 & 0 \\ 0 & 1 \end{vmatrix} & (1)\begin{vmatrix} 4 & 5 \\ 0 & 3 \end{vmatrix} \\ (-1)\begin{vmatrix} 0 & 3 \\ 3 & 1 \end{vmatrix} & (1)\begin{vmatrix} 1 & 3 \\ 0 & 1 \end{vmatrix} & (-1)\begin{vmatrix} 1 & 0 \\ 0 & 3 \end{vmatrix} \\ (1)\begin{vmatrix} 0 & 3 \\ 5 & 0 \end{vmatrix} & (-1)\begin{vmatrix} 1 & 3 \\ 4 & 0 \end{vmatrix} & (1)\begin{vmatrix} 1 & 0 \\ 4 & 5 \end{vmatrix} \end{bmatrix} = \begin{bmatrix} 5 & -4 & 12 \\ 9 & 1 & -3 \\ -15 & 12 & 5 \end{bmatrix}$$

Therefore the classical adjoint is then given by its transpose:

$$\text{adj } \mathbf{A} = (\text{cof } \mathbf{A})^T = \begin{bmatrix} 5 & -4 & 12 \\ 9 & 1 & -3 \\ -15 & 12 & 5 \end{bmatrix}^T = \begin{bmatrix} 5 & 9 & -15 \\ -4 & 1 & 12 \\ 12 & -3 & 5 \end{bmatrix}$$

The cofactors and classical adjoint are important because they are required in order to calculate the inverse of a matrix.

Theorem A.5. Matrix Inverse

If $\mathbf{A}$ is square and non-singular then $\mathbf{A}^{-1} = \dfrac{1}{\det \mathbf{A}} \text{adj } \mathbf{A}$

As an example, we will continue working with the same matrix $\mathbf{A}$ and find its inverse $\mathbf{A}^{-1}$. First we must calculate its determinant:

$$\det \mathbf{A} = \begin{vmatrix} 1 & 0 & 3 \\ 4 & 5 & 0 \\ 0 & 3 & 1 \end{vmatrix} = (1)\begin{vmatrix} 5 & 0 \\ 3 & 1 \end{vmatrix} - 0\begin{vmatrix} 4 & 0 \\ 0 & 1 \end{vmatrix} + 3\begin{vmatrix} 4 & 5 \\ 0 & 3 \end{vmatrix} = 41$$

Next, we use the formula for the inverse given in theorem A.5 and the adjoint matrix we calculated previously. Then we find that

$$\mathbf{A}^{-1} = \frac{1}{\det \mathbf{A}} \text{adj } \mathbf{A} = \frac{1}{41}\begin{bmatrix} 5 & 9 & -15 \\ -4 & 1 & 12 \\ 12 & -3 & 5 \end{bmatrix}$$

In practical terms, computation of the determinant is computationally inefficient, and there are faster ways to calculate the inverse, such as Gaussian Elimination. In fact, determinants and matrix inverses are very rarely used computationally because there is almost always a faster method in terms of number of computations required.

In R we can calculate the inverse using **solve** with a single argument. Normally **solve** is supplied two arguments and is used to solve a linear system. With a single argument, it returns the inverse of the square matrix.

```
M=matrix(c(1,0,3,4,5,0,0,3,1), nrow=3)
M
```

1	0	3
4	5	0
0	3	1

```
solve(M)
```

0.1219512	-0.09756098	0.29268293
0.2195122	0.02439024	-0.07317073
-0.3658537	0.29268293	0.12195122

Let **A** be an $n \times n$ square matrix and **0** be an $n \times 1$ vector of zeros, and suppose that
$$\mathbf{Av} = \mathbf{0}$$
for some vector $\mathbf{v} \neq \mathbf{0}$. Then it is not possible for **A** to have an inverse $\mathbf{A}^{-1}$, because if such an inverse were to exist, we would have
$$\mathbf{0} = \mathbf{A}^{-1}\mathbf{0} = \mathbf{A}^{-1}\mathbf{Av} = \mathbf{Iv} = \mathbf{v}$$

The same result follows if we multiply on the right. Together with theorem A.4, this proves the following theorem.

Theorem A.6. Singular Systems

If $\mathbf{v} \neq \mathbf{0}$ then
$$\mathbf{Av} = \mathbf{0} \iff \det \mathbf{A} = 0$$

Definition A.15. Eigenvalue/Eigenvector

Let **A** be a square matrix. Then we say that λ is an **eigenvalue** of **A** with corresponding **eigenvector** **v** if
$$\mathbf{Av} = \lambda \mathbf{v}$$
The pair $(\lambda, \mathbf{v})$ is called an **eigenvalue-eigenvector pair**.

We can re-write this by substituting an identity matrix,

$$\mathbf{Av} = \lambda \mathbf{Iv}$$

because $\mathbf{Iv} = \mathbf{v}$. Bringing everything to the left-hand side of the equation and right factoring the $\mathbf{v}$ gives

$$(\mathbf{A} - \lambda \mathbf{I})\mathbf{v} = \mathbf{0}$$

where $\mathbf{0}$ is an $n \times 1$ vector of zeros. Thus by theorem A.6, $\det(\mathbf{A} - \lambda \mathbf{I}) = 0$

Theorem A.7. Characteristic equation

Let $\mathbf{A}$ be a square matrix. Then its eigenvalues are the roots of its **characteristic equation***

$$\det(\mathbf{A} - \lambda \mathbf{I}) = \mathbf{0}$$

As an example, we will find the eigenvalues of the matrix

$$\mathbf{A} = \begin{bmatrix} 1 & 0 & 3 \\ 4 & 5 & 0 \\ 0 & 3 & 1 \end{bmatrix}$$

The characteristic equation is formed by subtracting λ from each diagonal element and setting the resulting determinant equal to zero.

$$0 = \begin{vmatrix} 1-\lambda & 0 & 3 \\ 4 & 5-\lambda & 0 \\ 0 & 3 & 1-\lambda \end{vmatrix} = 41 - 11\lambda + 7\lambda^2 - \lambda^3$$

The eigenvalues are $\lambda \approx 6.28761$, $\lambda \approx 0.356196 - 2.52861i$ and $\lambda \approx 0.356196 + 2.52861i$.

In R we there is a single function **eigen**.

```
M=matrix(c(1,0,3,4,5,0,0,3,1), nrow=3)
M
```

1	4	0
0	5	3
3	0	1

```
eigen(M)
```

```
eigen() decomposition
$values
[1] 6.287609+0.000000i 0.356196+2.528614i 0.356196-2.528614i

$vectors
              [,1]                    [,2]                    [,3]
[1,] -0.5707927+0i  -0.1488570+0.5846526i  -0.1488570-0.5846526i
[2,] -0.7545321+0i  -0.3456315-0.1882010i  -0.3456315+0.1882010i
[3,] -0.3238473+0i   0.6936441+0.0000000i   0.6936441+0.0000000i
```

Theorem A.8. Non-uniqueness of Eigenvectors

If **v** is an eigenvector of **A** with eigenvalue λ, then so is any multiple of **v**.

An example, consider the matrix

$$\mathbf{A} = \begin{bmatrix} 2 & -2 & 3 \\ 1 & 1 & 1 \\ 1 & 3 & -1 \end{bmatrix}$$

With some work we can find the characteristic equation. It is

$$0 = \begin{vmatrix} 2-\lambda & -2 & 3 \\ 1 & 1-\lambda & 1 \\ 1 & 3 & -1-\lambda \end{vmatrix}$$
$$= \cdots \text{ algebra omitted } \cdots$$
$$= -(\lambda+2)(\lambda-3)(\lambda-1)$$

Therefore the eigenvalues are -2, 3, 1. To find the eigenvector corresponding to -2 we must solve the following matrix equation for x, y, and z:

$$\begin{bmatrix} 2 & -2 & 3 \\ 1 & 1 & 1 \\ 1 & 3 & -1 \end{bmatrix} \begin{bmatrix} x \\ y \\ z \end{bmatrix} = -2 \begin{bmatrix} x \\ y \\ z \end{bmatrix}$$

This is equivalent to solving the following system of three equations in three unknowns x, y, and z:

$$2x - 2y + 3z = -2x$$

$$x + y + z = -2y$$
$$x + 3y - z = -2z$$

Rearranging by combining the terms on the right with those on the left,
$$4x - 2y + 3z = 0$$
$$x + 3y + z = 0$$
$$x + 3y + z = 0$$

The second and third equation are identical. This is because (as we were told in theorem A.8) the eigenvector is only determined in direction but not in magnitude. We are free to arbitrarily set any of x, y, and z. If we choose a nonzero value for $\mathbf{v}_j$ and it turns out that the remaining system has no solution, that means that $\mathbf{v}$ does not have a projection in the $\mathbf{e}_j$ direction and we should have chosen that component to be zero.

Suppose we try $y = 1$. The system becomes
$$4x - 2 + 3z = 0$$
$$x + 3 + z = 0$$

or moving the constants back to the right hand side of the equations,
$$4x + 3z = 2,$$
$$x + z = -3$$

Solving the second equation for $x = -z - 3$ and substituting into the first gives
$$4(-z - 3) + 3z = 2$$
which simplifies to $z = -14$. Thus $x = -z - 3 = -(-14) - 3 = 11$. An eigenvector corresponding to $\lambda = 2$ is
$$\mathbf{v} = [11, 1, -14]^{\mathrm{T}}$$
or any multiple thereof. A similar procedure can be used to find the two remaining eigenvectors.

Sometimes matrices will have a large number of zeros in them; if all of the elements of a square matrix fall on the main diagonal, the matrix is called a diagonal matrix. Diagonal matrices are particularly easy to work with because the eigenvalues can be read off and the determinant is the product of the diagonal elements.

Theorem A.9. Eigenvalues of a Diagonal Matrix

The eigenvalues of a diagonal matrix are the elements of the diagonal.

Even if a square matrix only has non zero elements on or above (or on or below) the main diagonal it is still very easy to work with, because the determinant is still the product of the diagonal elements.

Definition A.16. Triangular Matrices

An **upper (lower) triangular matrix** is a square matrix that only has its only nonzero entries on or above (below) the main diagonal.

Theorem A.10. Eigenvalues of Triangular Matrix

The eigenvalues of an upper (lower) triangular matrix are the numbers on the main diagonal.

Many efficient implementations of matrix operations (such as solving large systems of linear equations) involving reducing matrices to a triangular form.

The **Singular Value Decomposition** is a factoring of a matrix **X** into three parts, written as

$$\mathbf{X} = \mathbf{U}\Sigma\mathbf{V}^{\mathrm{T}} \qquad (A.1)$$

While all matrices have a singular value decomposition (SVD), when **X** is square, there is an intuitive description of the process: a rotation, followed by a scaling, followed by another rotation. In any decomposition, the first and third matrix are orthogonal, that is, they are composed of mutually perpendicular unit vectors; and the middle matrix is a diagonal matrix whose entries are called the **singular values** of **X**. This SVD is widely used in statistical and data analysis applications.

Theorem A.11. Singular Value Decomposition

Any $n \times m$ matrix $\mathbf{X}$ can be uniquely decomposed as a product

$$\mathbf{X} = \mathbf{U}\Sigma\mathbf{V}^\mathrm{T}$$

where $\mathbf{U}$ is an $n \times k$ orthogonal matrix, Σ is a $k \times k$ diagonal matrix, and $\mathbf{V}$ is an $m \times k$ orthogonal matrix.

The SVD has a geometric interpretation when $\mathbf{X}$ is a real, square 2×2 matrix. Think of the matrix $\mathbf{X}$ as an operator that transforms a vector $\mathbf{v}$, as in $\mathbf{v} \mapsto \mathbf{X}\mathbf{v}$. Then the SVD writes this transformation as

$$\mathbf{v} \mapsto \mathbf{X}\mathbf{v} = \mathbf{R}_2\mathbf{S}\mathbf{R}_1\mathbf{v}$$

where $\mathbf{R}_1$ and $\mathbf{R}_2$ are orthogonal matrices, and $\mathbf{S}$ is diagonal. Since $\mathbf{R}_1$ is orthogonal, it first rotates $\mathbf{v} \mapsto \mathbf{v}' = \mathbf{R}_1\mathbf{v}$ in a new coordinate system. Then the diagonal matrix expands (if $s_i > 1$) or compresses (if $s_i < 1$) the components of $\mathbf{v}'$ along the x' and y' axis. Finally, $\mathbf{R}_2$ is like an "unwinding" rotation, that brings the the expanded vector back, without changing its length.

Definition A.17. Singular Values

Let $\mathbf{X} = \mathbf{U}\Sigma\mathbf{V}^\mathrm{T}$ be the Singular Value Decomposition of $\mathbf{X}$. Then the values on the diagonal of Σ are the **singular values** of $\mathbf{X}$.

Definition A.18. Square Root of a Matrix

Let $\mathbf{A}$ be any matrix. Then its square root is any matrix $\sqrt{\mathbf{A}}$ such that

$$\mathbf{A} = (\sqrt{\mathbf{A}})(\sqrt{\mathbf{A}})$$

In general $\sqrt{\mathbf{A}}$ will not the matrix of square roots of $\mathbf{A}$.

Theorem A.12. Singular Value Decomposition Algorithm

Let $\mathbf{X}$ be any matrix over $\mathbb{R}$. Then $\exists$ orthogonal matrices $\mathbf{V}$ and $\mathbf{U}$ such that
$$\mathbf{X} = \mathbf{U}\Sigma\mathbf{V}^{\mathrm{T}}$$
here Σ is a diagonal matrix of singular values of $\mathbf{X}$, and in particular, for any orthonormal bases, $(\mathbf{v}_1, \ldots, \mathbf{v}_n)$ and $(\mathbf{u}_1, \ldots, \mathbf{u}_n)$, the matrices given by
$$\mathbf{U} = [\ \mathbf{u}_1\ |\ \ldots\ |\ \mathbf{u}_n\ ]$$
$$\mathbf{V} = [\ \mathbf{v}_1\ |\ \ldots\ |\ \mathbf{v}_n\ ]$$
where the $\mathbf{v}_j$ are normalized eigenvectors of $\mathbf{S} = \sqrt{\mathbf{X}^{\mathrm{T}}\mathbf{X}}$ and $\mathbf{u}_j = \Sigma\mathbf{v}_j$, where
$$\mathbf{X} = \Sigma\mathbf{S} = \Sigma\sqrt{\mathbf{X}^{\mathrm{T}}\mathbf{X}}$$
and the elements of Σ are the square roots of the eigenvalues of $\mathbf{S}$.

Algorithm A.1 describes how to calculate the singular value decomposition.

Algorithm A.1 Algorithm for the singular value decomposition.

---

input: M
1: $\mathbf{v}_1, \mathbf{v}_2, \ldots \leftarrow$ Eigenvectors $(\mathbf{M}^{\mathrm{T}}\mathbf{M})$
2: $\mathbf{u}_1, \mathbf{u}_2, \ldots \leftarrow$ Eigenvectors $(\mathbf{M}\mathbf{M}^{\mathrm{T}})$
3: $\mathbf{S} \leftarrow$ Diagonal(Eigenvalues $(\mathbf{M}^{\mathrm{T}}\mathbf{M}))$
4: $\mathbf{V} \leftarrow [\ \mathbf{v}_1^{\mathrm{T}}\ |\ \mathbf{v}_2^{\mathrm{T}}\ |\ \cdots\ ]$
5: $\mathbf{U} \leftarrow [\ \mathbf{u}_1^{\mathrm{T}}\ |\ \mathbf{u}_2^{\mathrm{T}}\ |\ \cdots\ ]$
6: **repeat**
7:    **if** $\mathbf{U}\mathbf{S}\mathbf{V}^{\mathrm{T}} = \mathbf{M}$ **then**
8:       **return** $(\mathbf{U}, \mathbf{S}, \mathbf{V})$
9:    **end if**
      Flip the direction of one of the column vectors of $\mathbf{U}$ or $\mathbf{V}$
10: **until** all permutations have been checked.

The following result has significant applications in many fields of analysis, and is particularly useful for image compression.

Theorem A.13. SVD Expansion

The SVD can be expanded as

$$\mathbf{X} = \mathbf{U\Sigma V}^T = s_1\mathbf{u}_1\mathbf{v}_1^T + s_2\mathbf{u}_2\mathbf{v}_2^T + \cdots + s_n\mathbf{u}_n\mathbf{v}_n^T$$

where $\mathbf{U}$ and $\mathbf{V}$ are as previously defined and the the u_i and v_i and the column and row vectors of $\mathbf{U}$ and $\mathbf{V}$.

The SVD can be found in R with the function **svd**

```
M=matrix(c(1,0,3,4,5,0,0,3,1), nrow=3)
M
```

1	4	0
0	5	3
3	0	1

```
svd(M)
```

```
$d
6.90579957445495 3.10831938531505 1.9100478623146
$u
0.5501849      -0.05784255    -0.8330371
0.8278080       0.16881578     0.5350094
0.1096835      -0.98394892     0.1407624
$v
0.1273184      -0.9682690     -0.2150469
0.9180370       0.1971190     -0.3440235
0.3754971      -0.1536205      0.9140037
```

The individual matrices **d**, **u**, and **v** can be extracted using the dollar-sign operator, e.g.,

```
mysvd=SVD(M)
d=mysvd$d
u=mysvd$u
v=mysvd$v
```

General References

[ADLER] Adler, J. (2012) **R in a Nutshell. 2nd Ed.** O'Reilly. All levels. Programming in R, with a focus on data mining and machine learning rather than statistics. Like other books in the *Nutshell* series.

[ALPAY] Alpaydin, E. (2014) **Introduction to Machine Learning. 3rd Ed.** MIT Press. Mathematics of ML, undergraduate.

[BISHOP] Bishop, C. (2006). **Pattern recognition and machine learning.** Springer. Standard reference on the subject.

[ESL] Hastie, T., Tibshirani, R., Friedman, J. (2009) **The Elements of Statistical Learning. Data Mining, Inference, and Prediction. 2nd Ed.** Springer. Full Text available online at https://web.stanford.edu/~hastie/Papers/ESLII.pdf. This is a more advanced version of [ISL] without the code. Mathematical. Advanced undergraduate/graduate.

[ISL] James, G., Witten, D., Hastie, T., Tibshirani, R. (2013) **An Introduction to Statistical Learning, with Applications in R**. Springer. Full text available online at http://www-bcf.usc.edu/~gareth/ISL/. Undergraduate, all levels. This is a more accessible version of [ESL], with a lot of R.

[MARSL] Marsland, S. (2015) **Machine Learning, An Algorithmic Perspective. 2nd Ed.** CRC Press. All levels. Includes math, algorithms, and code in Python. Where code is given, it does not show you how to use high level machine learning packages in Python, rather it shows you how to design your own implementations. If you don't know Python but want to learn the math and algorithms this is a good place to start - you can just skip the code blocks.

[MASS] Venables, W., Ripley, D. (2002) **Modern Applied Statistics with R. 4th Ed.** Springer. This book is mainly of interest because of the large number of data sets from it that were contributed to CRAN and are now open source. The book itself is expensive and rarely used but you can probably find an un-indexed pdf file if you look around.

[MURPHY] Murphy, K. (2012) **Machine Learning. A Probabilistic Perspective**. MIT Press. Advanced. Mathematical. Perspective of theoretical statistics.

[WACKER] Wackerly, D., Mendenhall, W., Scheaffer, R.. (2008) **Mathematical statistics with applications. 7th Ed.** . Thomson Brooks/Cole. Standard undergraduate mathematical statistics textbook.

[ZAKI] Zaki, M., Meira, W. (2014) **Data Mining and Analysis. Fundamental Concepts and Algorithms.** Cambridge University Press. Advanced undergraduate. Mathematics and algorithms. Complete text is available for download from http://www.cs.rpi.edu/~zaki/dataminingbook/pmwiki.php.

Index

abline, 39
adabag, 172
agnes, 197, 198
apply, 76, 154
argmin, 180
array_reshape, 117
arrays, 19
arrows, 152
as.*datatype*, 16
as.integer, 223
atomic data types, 16
AUC, 114
available.packages, 14

B, 231
backpropagation, 73
bagging, 169
basis vectors, 251
biocLite, 15
bioconductor, 15
blocks, 27
boosting, 172
boosting, 173
bootstrap aggregation, 169
boundary extraction, 243
break, 26

c (combine), 16
Canny filter, 241
caret, 109, 136
CART, 87
cat, 23
chapter-naive-bayes, 131
chapter-SOM, 212
characteristic equation, 261
cimg, 233
circularize_dendrogram, 199
class, 127
classvec2classmat, 215
closure, 251
cluster, 198
CMY, 229
cofactor, 257
colMeans, 151
colnames, 24, 34, 79

color_branches, 201
column width in Jupyter, 223
confusionMatrix, 109, 136, 167
contrast stretching, 232
cut point, 163
cutree, 203

data frame, 34
data frames, 23
data.frame, 23
dbscan, 205
decision tree, 163
dendrogram, 196
det, 255
determinant, 254, 256
diagonal matrix, 263
dilation (image), 242
dot product, 250
dropout (keras), 121

e1071, 132, 161
ecdf, 233
edge detection, 237
eigen, 261
eigenvalue, 260, 262
eigenvector, 260, 262
else, 27
energy function, 42
ensemble learning, 169
entropy, 163
erosion (image), 242
evaluate (keras), 125

F1 score, 109
fallout, 108
false positive rate, 108
family (**glm**), 62
feature matrix, 147
FM_index, 204
for, 24
format, 80

G, 231
glm, 58, 61
glmnet, 65

`grayscale`, 231

Hacker's Codes, iv
`head`, 18, 223
Hebbian learning rule, 221
hierarchical clustering, 193
`hist`, 103, 233
homunculus, 212, 213

identifiers, 15
`if`, 27
`if`, nested, 28
`image`, 154
image convolution, 235
image smoothing, 236
`image_convolve`, 240
`image_negate`, 240
`image_read`, 239
`imager`, 230
`imsub`, 231
`install.packages`, 14
`intTobits`, 223
`ipred`, 170
`is.`*datatype*, 16

`jpeg`, 152

`kmeans`, 183
`knn`, 127
`kohonen`, 214

Laplacian Filter, 238, 239
lasso regression, 65
LDA, 138
`lda`, 141
least squares, 43, 44, 47, 48
`library`, 14
linear dependence, 251
linear discriminant analysis, 138
linear regression, 43, 44
linear spatial filtering, 234
`lm`, 36
`lm.ridge`, 65
`load.image`, 230, 239
logistic regression, 100
logit function, 100
loops in R, 24
loss function, 122

`magick`, 230
markdown, 13
matrices, 19
matrix, 252
matrix addition, 253
matrix adjoint, 258
matrix identity, 257
matrix inverse, 257, 259
matrix multiplication, 256
matrix transpose, 252
metric (keras), 122
MSS error, 40
multilinear regression, 57, 63
`mvrnorm`, 127, 188

`na.omit`, 58, 75
`naiveBayes`, 135
`neuralnet`, 75, 77
`nls`, 68
normal equations, 48
normal matrix, 48
`nrow`, 36

objective function, 42, 47
one-hot encoding, 118
optimizer, 122
output of a cell, 13

`pairs`, 61
`paste`, 22, 172
PCA, 145
perceptron, 70
piping, 240
`plot`, 35
`plotcop`, 94
`plyr`, 103
polynomial least squares, 47, 48
`prcomp`, 151
`predict`, 41, 95, 105, 126, 165, 170, 178
principal components, 148
`printcp`, 94
`pROC`, 113, 137

QDA, 142
`qda`, 143
Quadratic Discriminant Analysis, 142

`R`, 231

R squared, 40
random forest, 176
random forests, 176
random.colors, 207
rbind, 79
read.table, 34
readJPEG, 152
recall, 108
regression tree, 87
rep, 185
repeat, 26
rev, 154
revalue, 103
reverse, 223
RGB, 228
rgb, 184
ridge regression, 64
roc, 113, 137
ROC curve, 111
row.names, 24
rpart, 92, 165
rpart.control, 92, 93
rpart.plot, 165
RSS error, 40
runif, 184

sample, 30, 36
sapply, 28
save.image, 239
scalar product, 251
scale, 76, 215
scatter matrix, 147, 148
segmentation, 245
sensitivity, 108
seq, 16
set.my.colors, 207
set.random.colors, 208
set.seed, 31
sil.score, 186
silhouette index, 186
singular linear system, 260
singular matrix, 255
singular value decomposition, 265

singular values, 265
slices (of vectors), 18
Sobel filter, 241
sofmax, 101
som, 216
sort, 83
source, 14
specificity, 108
splitting criteria, 163
sprintf, 23
square root of a matrix, 265
subset, 34, 79
sum, 34, 41
summary(keras), 121, 122
summary, 38
supervised learning, 2
SVD, 267
svm, 161

t, 154, 253
table, 167
tail, 18
tanglegram, 202
thresholding, 236
to_categorical, 119
topological clustering, 205
tree, 90
true positive rate, 108
typeof, 16

unsupervised learning, 3

variance-bias tradeoff, 55
vector, 16, 248
vector addition, 250
vector magnitude, 249

while, 25
width, 18

xyf, 218

zero centering, 147

Thank you for buying my book. I hope you have enjoyed reading it as much as I have enjoyed writing it.

If you have any comments or suggestions I would love to hear from you. Please feel free to write to me at `ibellaromeo@gmail.com`.

Bella

About the Author

Isabella Romeo is a pseudonym for Bruce E. Shapiro. Bella is a blue nose American Staffordshire Terrier. Romeo, a red nose pit, was her life partner and constant companion. Bella's hobbies include barking at trees, chasing grasshoppers, and sleeping on the living room couch. Since she does not have opposable thumbs, she dictated this book to her human, who, in turn, did all the typing for her. Not only does he have opposable thumbs, sometimes it seems like he has ten of them (for example, he totally ignores the little red lines under words - wtf you say? spell check?).

Figure A.1.: Bella, always eager for a snack, is on the right. Her late brother Romeo is on the left.

Said human with opposable thumbs and extraordinary typing skils has taught undergraduate mathematics at a California State University campus for the past two decades. In his alternate and sometime prior incarnations he was a senior researcher in the machine learning group at JPL, a computational scientist at Caltech, and a satellite orbital design and mission engineer under contract at NASA/Goddard Space Flight Center.

Printed in Great Britain
by Amazon